Advances in Intelligent Systems and Computing

Volume 271

Series editor

Janusz Kacprzyk, Polish Academy of Sciences, Warsaw, Poland
e-mail: kacprzyk@ibspan.waw.pl

For further volumes:
http://www.springer.com/series/11156

About this Series

The series "Advances in Intelligent Systems and Computing" contains publications on theory, applications, and design methods of Intelligent Systems and Intelligent Computing. Virtually all disciplines such as engineering, natural sciences, computer and information science, ICT, economics, business, e-commerce, environment, healthcare, life science are covered. The list of topics spans all the areas of modern intelligent systems and computing.

The publications within "Advances in Intelligent Systems and Computing" are primarily textbooks and proceedings of important conferences, symposia and congresses. They cover significant recent developments in the field, both of a foundational and applicable character. An important characteristic feature of the series is the short publication time and world-wide distribution. This permits a rapid and broad dissemination of research results.

Advisory Board

Keon Myung Lee · Seung-Jong Park
Jee-Hyong Lee
Editors

Soft Computing in Big Data Processing

Springer

Editors
Keon Myung Lee
Chungbuk National University
Cheongju
Korea

Jee-Hyong Lee
Sungkyunkwan University
Gyeonggi-do
Korea

Seung-Jong Park
Louisiana State University
Louisiana
USA

ISSN 2194-5357 ISSN 2194-5365 (electronic)
ISBN 978-3-319-05526-8 ISBN 978-3-319-05527-5 (eBook)
DOI 10.1007/978-3-319-05527-5
Springer Cham Heidelberg New York Dordrecht London

Library of Congress Control Number: 2014933211

Printed on acid-free paper

Springer is part of Springer Science+Business Media (www.springer.com)

Preface

Big data is an essential key to build a smart world as a meaning of the streaming, continuous integration of large volume and high velocity data covering from all sources to final destinations. The big data range from data mining, data analysis and decision making, by drawing statistical rules and mathematical patterns through systematical or automatically reasoning. The big data helps serve our life better, clarify our future and deliver greater value. We can discover how to capture and analyze data. Readers will be guided to processing system integrity and implementing intelligent systems. With intelligent systems, we deal with the fundamental data management and visualization challenges in effective management of dynamic and large-scale data, and efficient processing of real-time and spatio-temporal data. Advanced intelligent systems have led to managing the data monitoring, data processing and decision-making in realistic and effective way. Considering a big size of data, variety of data and frequent changes of data, the intelligent systems basically challenge new data management tasks for integration, visualization, querying and analysis. Connected with powerful data analysis, the intelligent systems will provide a paradigm shift from conventional store and process systems. This book focuses on taking a full advantage of big data and intelligent systems processing. It consists of 11 contributions that feature extraction of minority opinion, method for reusing an application, assessment of scientific and innovative projects, multi-voxel pattern analysis, exploiting No-SQL DB, materialized view, TF-IDF criterion, latent Dirichlet allocation, technology forecasting, small world network, and classification & regression tree structure. This edition is published in original, peer reviewed contributions covering from initial design to final prototypes and authorization.

To help readers understand articles, we describe the short introduction of each article as follows;

1. "Extraction of Minority Opinion Based on Peculiarity in a Semantic Space Constructed of Free Writing": This article proposes a method for extracting minority opinions from a huge quantity of text data taken from free writing in user reviews of products and services. The extraction becomes outliers in a low-dimensional semantic space. Peculiarity Factor (PF) enables them to extract minority opinions for outlier detection.

2. "System and Its Method for Reusing an Application": This article describes the system that can reuse the pre-registered applications in the open USN (Ubiquitous Sen-

sor Network) service platform. Cost and time can be reduced for implementing a duplicated application.

3. "Assessment of Scientific and Innovative Projects: Mathematical Methods and Models for Management Decisions": This article is designed to improve the quality of assessment of scientific and innovative projects the mathematical methods and models. This methodology is applied to expert commissions who are responsible for venture funds, development institutes, and other potential investors required selecting appropriate scientific and innovative projects.

4. "Multi-voxel pattern analysis of fMRI based on deep learning methods": This paper describes constructing a decoding process for fMRI data based on Multi-Voxel Pattern Analysis (MVPA) using deep learning method for online training process. The constructed process with Deep Brief Network (DBN) extracts the feature for classification on each ROI of input fMRI data.

5. "Exploiting No-SQL DB for Implementing Lifelog Mashup Platform": To support effcient integration of heterogeneous lifelog service, this article states exploiting the lifelog mashup platform with the document-oriented No-SQL database MongoDB for the LLCDM repository. It develops an application of retrieving Twitter's posts involving URLs.

6. "Using Materialized View as a Service of Scallop4SC for Smart City Application Services": This paper proposes materialized view to be as service (MVaaS). A developer of an application can efficiently and dynamically use large-scale data from smart city by describing simple data specification without considering distributed processes and materialized views. It designs an architecture of MVaaS using MapReduce on Hadoop and HBase KVS.

7. "Noise Removal Using TF-IDF criterion for Extracting Patent Keyword": This article proposes a new criteria for removing noises more effectively and visualizing the resulting keywords derived from patent data using social network analysis (SNA). It can quantitatively analyze patent data using text mining with TF-IDF used as weights and classify keywords and noises by using TF-IDF weighting.

8. "Technology Analysis from Patent Data using Latent Dirichlet Allocation": This paper discusses how to apply latent Dirichlet allocation, a topic model, in a trend analysis methodology that exploits patent information. To perform this study, they use text mining for converting unstructured patent documents into structured data, and the term frequency-inverse document frequency (tfidf) value in the feature selection process.

9. "A Novel Method for Technology Forecasting Based on Patent Documents": This paper proposes a quantitative emerging technology forecasting model. The contributors apply patent data with substantial technology information to a quantitative analysis. They can derive a Patent–Keyword matrix using text mining that leads to reducing its dimensionality and deriving a Patent–Principal Component matrix. It makes a group of the patents altogether based on their technology similarities using the K-medoids algorithm.

10. "A Small World Network for Technological Relationship in Patent Analysis": Small world network consists of nodes and edges. Nodes are connected to small steps of edges. This article describes the technologies relationship based on small world network such as the human connection.

11. "Key IPC Codes Extraction Using Classification and Regression Tree Structure": This article proposes a method to extract key IPC codes representing entire patents by using classification tree structure. To verify the proposed model to improve performance, a case study demonstrates patent data retrieved from patent databases in the world.

We would appreciate it if readers could get useful information from the articles and contribute to creating innovative and novel concept or theory. Thank you.

Keon Myung Lee
Seung-Jong Park
Jee-Hyong Lee

Contents

Extraction of Minority Opinion Based on Peculiarity in a Semantic Space Constructed of Free Writing
Analysis of Online Customer Reviews as an Example

Fumiaki Saitoh, Takuya Mogawa, and Syohei Ishuzu

Department of Industrial and Systems Engineering,
College of Science and Engineering,Aoyama Gakuin University
5-10-1 Fuchinobe,Chuo-ku,Sagamihara City, Kanagawa, Japan
saitoh@ise.aoyama.ac.jp

Abstract. Recently, the "voice of the customer (VOC)" such as exhibited by a Web user review has become easily collectable as text data. Based on large quantity of collected review data, we take the stance that minority review sentences buried in a majority review have high value. It is conceivable that hints of solution and discovery of new subjects are hidden in such minority opinions. The purpose of this research is to extract minority opinion from a huge quantity of text data taken from free writing in user reviews of products and services. In this study, we propose a method for extracting minority opinions that become outliers in a low-dimensional semantic space. Here, a low-dimensional semantic space of Web user reviews is constructed by latent semantic indexing (LSI). We were able to extract minority opinions using the Peculiarity Factor (PF) for outlier detection. We confirmed the validity of our proposal through an analysis using the user reviews of the EC site.

Keywords: Text Mining, Online Reviews, Peculiarity Factor (PF), Dimensionality Reduction, Voice of the Customer (VOC).

1 Introduction

The purpose of this research is to extract minority opinion from a huge quantity of text data taken from free writing in user reviews of products and services. Because of the significant progress in communication techniques in recent years, the "voice of the customer (VOC)" such as exhibited by Web user reviews has become easily collectable as text data. It is considered important to use VOC contributions to improve customer satisfaction and the quality of services and products[1]-[2].

In general, although text data of Web user reviews are readily available, they are typical concentrated into "satisfaction" or "dissatisfaction" categories. Existing text-mining tools can extract knowledge about such a majority user review sentence easily. Furthermore, because there is a strong possibility that

K.M. Lee, S.-J. Park, J.-H. Lee (eds.), *Soft Computing in Big Data Processing*,
Advances in Intelligent Systems and Computing 271,
DOI: 10.1007/978-3-319-05527-5_1, © Springer International Publishing Switzerland 2014

knowledge about a majority review is obvious knowledge and an expected opinion, extraction of useful knowledge such as a new subject or hints of a solution can be difficult. For example, dissatisfied opinion concentrates on staff behavior in a restaurant where service is bad, and many satisfactory opinions about image quality are acquired for a product with a high-resolution display. It is easy to imagine concentrations of similar opinions occurring frequently as these examples.

Based on a large quantity of collected review data, we take the stance that minority review sentences buried in a majority review have high value. We aim at establishing a method for extracting minority Web review opinions.

Because Web review sentences consist of free writing, the variety of words appearing can be extensive. Word frequency vectors become very large sparse vectors that have many zero components. Therefore, an approach based on the frequency of appearance of words is not suitable for this task. However, approaches using TF-IDF as an index can be used for extracting characteristic words, but because these indexes function effectively only on data within large documents, they are unsuitable for Web review sentences, which appear within documents.

Therefore, we define a minority opinion Web review as an outlier in the semantic space of a Web review. In this study, we extract minority Web reviews through outlier detection on a semantic space constructed by using latent semantic indexing (LSI). The merit of our proposal is that it allaws us to extract minority opinion from free writing reviews accumulated in large quantities, without the need to read these texts.

2 Technical Elements of the Proposed Method

In this section, we outline the technical elements of our proposed method.

2.1 Morphological Analysis

To analyze a Web review comprising qualitative data quantitatively, it is necessary to extract the frequency of appearance of the words in each document as a feature quantity of the document. The number of lexemes appearing in all the data is a component of the word frequency vector. Morphological analysis is a technique for writing with segmentation markers between lexemes that is the unit of a vocabulary. This natural language processing is carried out based on information about grammar and dictionaries for every language[3][4].

Because the word frequency vector is measured easily, this processing is unnecessary for languages segmented at every word, such as English. However, it is a very useful tool for text mining of unsegmented languages such as Japanese, Korean and Chinese. We created a word frequency vector through a morphological analysis, because our subject was Japanese Web reviews.

2.2 Latent Semantic Indexing

Latent Semantic Indexing (LSI)[5] is well-known method for natural language processing based on mathematical techniques. To effectively detect outliers used for extraction of minority opinions, we reduct the dimension of the word frequency vector by using LSI. The word frequency vector, which is a large sparse vector, is transformed into a low-dimensional dense vector by using this technique, while preserving semantic structure.

Low-dimensional semantic space data are obtaind by performing singular value decomposition (SVD) on the following document matrix whose column vector is a word frequency vector:

$$D = U\Sigma V^T, \tag{1}$$

$$\Sigma = \left(\begin{array}{cccc|c} \sigma_1 & 0 & \dots & 0 & \\ 0 & \sigma_2 & & \vdots & \\ \vdots & & \ddots & 0 & O_{r\times(n-r)} \\ 0 & \dots & 0 & \sigma_r & \\ \hline & O_{(m-r)\times r} & & & O_{(m-r)\times(n-r)} \end{array}\right), \tag{2}$$

where, U is an $m{\times}m$ unitary matrix, Σ is an $m{\times}n$ rectangular diagonal matrix, and V^T is the conjugate transpose matrix of V which is an $n{\times}n$ unitary matrix. The diagonal entries $\sigma_i(\sigma_1 \geq \sigma_2, \dots, \geq \sigma_r > 0, i = 1, 2, \dots, r)$ of Σ are known as the singular value of D and are arranged in order from small to large values. r is the rank of matrix Σ and all the entries of the matrix Σ are zero except σ_i. $O_{(\cdot)\times(\cdot)}$ is the null submatrix.

3 Proposed Method

In this section, we explain the details of our proposal.

3.1 Basic Policy of the Proposal

Review sentences that express similar opinions have a strong tendency to include identical or similar words. In the low-dimensional semantic space generated by using LSI, data from like reviews are located in the same neighborhood. When much data of similar opinion exist, they become clustered in semantic space. However, because data without similar data cannot constitute a cluster, they become outliers in semantic space.

A schematic diagram depicting the relation between minority opinion and majority opinion in semantic space is shown in Fig.1. Shown is an example of the opinion for a cafe. For restaurants, it is easy to predict that customer (dis)satisfaction and opinion concentrate on quality of service and food.

In contrast, it is not easy to predict a request for a power supply for using a notebook PC, in a customer review. There is every possibility that data from such an opinion will become an outlier in semantic space. The goal of this research is to detect such a minority group review quantitatively.

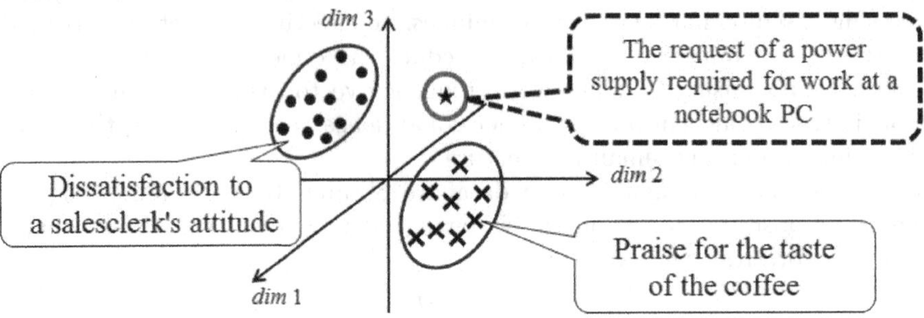

Fig. 1. Schematic diagram of semantic space based on review data constructed by using a LSI

3.2 Extraction of Minority Opinion Using the Peculiarity Factor

Here we explain the proposed method for extracting a minority opinion defined asan outlier in semantic space constructed by techniques that will be introduced in the next section. In this study, we use the peculiarity factor (PF) for outlier detection in a multidimensional data space. The peculiarity factor (PF) is a criterion for evaluating the peculiarity of data and is widely used for outlier detection[6][7]. PF values are calculated based on the distance between each datum and are calculated for each datum.

PF value become large when the datum for calculation is separated from the cluster of other data. By using this disposition, we can recognize data with extremely large PF values as outlier data. The merits of using the peculiarity factor are that a user does not assume a data distribution beforehand and that a user can understand the distribution visually.

The attribute $PF(PF_a)$ is used for single-variable data and the record $PF(PF_r)$ that is used for multivariable data. The definition of PF_a and PF_r are explained as follows. Let, $X=\{\boldsymbol{x}_1, \boldsymbol{x}_2, \dots, \boldsymbol{x}_{n-1}, \boldsymbol{x}_n\}$ be n sampled dataset with $\boldsymbol{x}_i = (x_{i1}, x_{i2}, \dots, x_{im})$. For the dataset X, PF_a of $\boldsymbol{x}_{ij}$ is calculated by

$$PF_a(\boldsymbol{x}_{ij}) = \sum_{j=1}^{n} N(x_{ij}, x_{kj})^{\alpha}, \tag{3}$$

$$N(x_{ij}, x_{kj}) = \|x_{ij} - x_{kj}\| \tag{4}$$

where n is the sample size. α is a parameter, usually set to 1. PF_r is calculated by substituting PF_a in

$$PF_r(\boldsymbol{x}_i) = \sum_{j=1}^{n} \left(\sum_{l=1}^{m} \beta \times (PF_a(x_{il}) - PF_a(x_{ij}))^q \right)^{\frac{1}{q}}, \tag{5}$$

where, m is the number of data dimensions, β is the weight parameter for each attribute, q is a degree parameter, and $PF_a(x_{ij})$ is given by Eq.(3).

Minority topics are extracted based on the resulting calculation of PF_r. The extraction method of a minority opinion is as follows. Let, $Y=\{y_1, y_2, ..., y_{n-1}, y_n\}$ be n sampled label sets for dataset X. These labels are assigned based on the following criterion:

$$y_i = \begin{cases} 1 \; if(PF_r(\boldsymbol{x}_i) > \theta) \\ 0 \; otherwise \end{cases} \tag{6}$$

where θ is the threshold of the PF value for identifying minorities and others. This value is determined based on the user's heuristics, since data with extremely large PF values can easily confirmed by sight inspection of a graph.

3.3 Proposed Method Procedure

The procedure for our proposed method is as follows:

Step.0 The text data set of analysis subjects is written with segmentation markers between lexemes based on morphological analysisx.(For segmented languages such as English, step.0 is skipped.)
Step.1 The component of a word vector is determined by using the words found in all the review sentences.
Step.2 the word frequency vector is prepared by counting the number of appearances of words.
Step.3 The word frequency vector, which is a large sparse vector, is transformed into a low-dimensional dense vector through LSI.
Step.4 The PF value is calculated for each data sets using Eq.(4).
Step.5 Review sentences with extremely large PF values as calculated in Step 4 are searched for. Any review sentence regarded as an outlier is extracted as a minority opinion.

It is easy to discover minority opinions by visualizing the PF values using, e.g. bar chart.

4 Experiment

Here we explain the details of experiment.

4.1 Experimental Settings

If the data consist of review sentences of a free description type, our proposed technique is applicable to any objects, such as bulletin board systems (BBSs), questionnaires and online review sites. We adopted Web user reviews as an example of data for verification of our proposal. Such reviews send the voice of the customer to the analyst directly and can make a significant contribution to quality and customer satisfaction.

Because of the rapidly expanding market for tablet PCs, Web review data about certain tablet PCs were selected as the subject of our analysis. We collected review sentences written on certain well-knows EC sites as candidates for analysis. The language analyze was limited to Japanese. Because some English and Chinese were also contained in the data set and these would become noise, they were removed beforehand.

Not, all review sentences are necessarily equivalent in length. Reviewers' earnestness to the product and their nature can affect the length of each review sentence. Hence, we investigated the number of characters per data as the length of review sentences. Fig.2 shows the distribution of the number of characters. The vertical axis shows the number of characters and the horizontal axis shows the frequency. Although the frequency decreases with increasing number of characters, review sentences that have numerous characters do exist slightly.

Regardless of the content of the text, such lengthy revews are detected as outliers by using the proposed method in our preliminary experiment. This is because of the abundance and the kind and frequency of the words used in extremely long review sentences. Thus, the data set was divided according to the number of characters of review data, so as to eliminate ill effects from the difference in the number of characters.

The data were analyzed after being divided into three sets based on review data whose numbers of characters are 100 or fewer, because these occupy most of the frequency spectrum. The parts of speech selected as components of a word frequency vector are nouns and adjectives. The Parameter set shown in table.1 was used for this experiment. These values are determined based on users' heuristics.

Table 1. List of parameters

Parameters for PF	
α	1
β	1
q	2
Parameters for LSI	
Dimension of U	10
Dimension of V^T	10
The threshold value θ	
θ for data set whose number of characters are 100 or less (fig.3)	300

4.2 Experimental Result

The experimental results based on calculation of the PF value are shown in the Fig.3. These bar graphs have vertical axes showing the PF value and horizontal axes showing the number of data. Examples of review data results of minority opinion extraction using the proposed method are illustrated in Table.2 Table.3 shows examples of review data identified as majority opinion. (Since a review sentence cannot be published directly, these examples express reviews in slightly

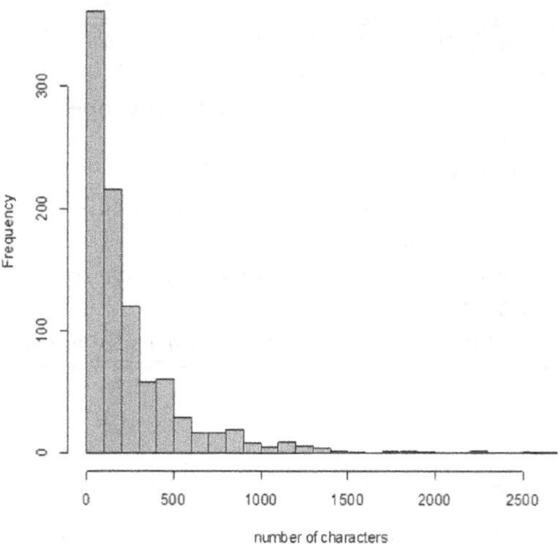

Fig. 2. Distribution of the number of characters used for the review sentences

rewritten form). In addition, proper nouns are not published to avoid specification of a product name or maker. In Fig.3, the existence of data with extremely large PF values can be confirmed by inspection.

4.3 Discussion

We compared majority reviews and minority reviews extracted by using our proposal. The validity of our proposal can be confirmed by the difference in content between minority reviews and majority reviews (see Table.2). However, minority opinions could not necessarily be extracted perfectly by using the proposed

Table 2. Examples of review sentences extracted as majority opinions

⋆ Although it won in ∘∘ as compared with the product X of A comp. or the product Y of B comp., in ••, it is inferior.
⋆ The screen resolution is very clear, and the speaker sound is great.
⋆ Wi-Fi reception is very good and it runs smoothly because touchscreen has high responsiveness.
⋆ Display on the home screen is very easy to understand and I can operate it intuitively.
⋆ A hand will get tired if it is used for a long time, since the weight of a main part is a little heavy.
⋆ Since it does not correspond to service of ♣♣, only ♢♢ can be used.
⋆ Although I wanted to play △△ (game application that is very popular in japan), it did not correspond to △△.

Table 3. Examples of review sentences extracted as minority opinions:for data set whose number of characters are 100 or less

⋆ It is hard for elderly people to use it without a polite description. I want the consideration to elderly people.
⋆ Dissatisfaction is at the point which has restriction in cooperation with DDD.
⋆ It doesn't have a SD/microSD card slot. The lack of a SD card slot is a mistake.
⋆ Since it corresponds only to application of *A* company, ○○ (application of *B* company) is not available.
⋆ Although I heard the rumor that ◇◇ could not use, I have solved this problem by installing ♡♡ and ♠♠. I'm satisfied.
⋆ I used it as alarm clock, but it shut downed automatically and did not sound.

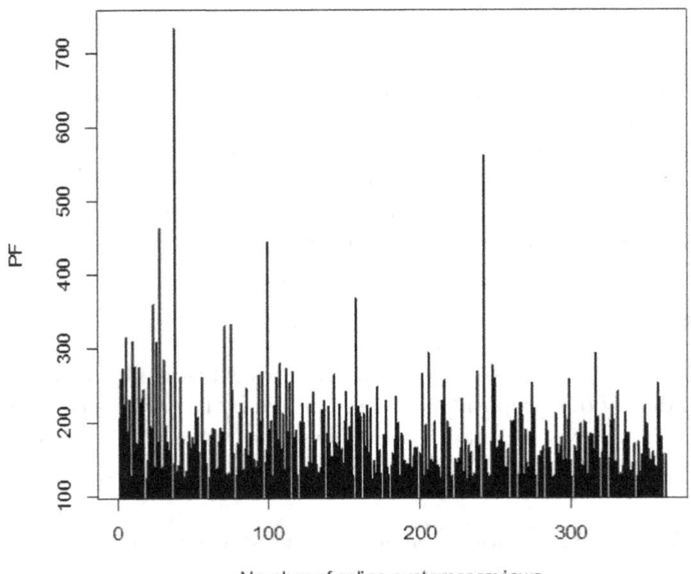

Fig. 3. The PF value about the review data whose number of characters are 100 or less

method, and the reviews can be interpreted as a majority opinions were mixed slightly into the extraction result.

We considered that this was due to the difference in the mode - of expression (for example, the existence or nonexistence of picture language, personal habit of wording, and so on) of each reviewer. The value of the extracted minority opinion depends on knowledge that the user has previously. It is possible that ordinary opinion and obvious knowledge are contained in extracted results. Hence, even if extraction results are minority opinions, these are not always worthy.

However, despite these shortcomings we facilitated extraction of worthy knowledge through visualization of opinion peculiarity based on our proposal thereby making it easier to reflect important opinions that tended to be disregarded until now in the design of a product or service.

We were unable to provide a sufficient argument about the computational complexity of the PF nor about the relation between PF value and extraction accuracy. These are far more complicated problems and need further study. In addition, the possibility that the length of these reviews follows a power-law distribution is confirmed through visual observation of the histogram[8]. Since there are a few extremely long sentences, this figure suggests that the length of a sentence has a long-tailed distribution. We will advance our investigation into this point in the future.

5 Conclusions

In this study, we proposed a method for extracting minority opinion from a huge quantity of customer reviews, from the viewpoint that data about minority opinion contain valuable knowledge. Here we defined a minority opinion as outlier customer review data in semantic space acquired by using LSI and extracted these data based on their PF value. We confirmed the validity of our proposal through experiments using Web user reviews collected from the EC site. It was confirmed that the content of user reviews extracted as minority opinions cleary differed from that of majority opinions. We successfully extracted buried knowledge, without having to read each user sentence.

In addition, our investigation about length of review sentence, of review sentence length suggests that this length follows a power-law distribution. We expect that techniques such as ours will be further developed in application to customers' free writing, since the number of customer voices on the Web is going to further increase.

References

1. Jain, S.S., Meshram, B.B., Singh, M.: Voice of customer analysis using parallel association rule mining. In: IEEE Students' Conference on Electrical Electronics and Computer Science (SCEECS), pp. 1–5 (2012)
2. Peng, W., Sun, T., Revankar, S.: Mining the "Voice of the Customer" for Business Prioritization. ACM Transactions on Intelligent Systems and Technology 3(2), Article 38 (2012)
3. Yamashita, T., Matsumoto, Y.: Language independent morphological analysis. In: Proceedings of the Sixth Conference on Applied Natural Language Processing (ANLC 2000), pp. 232–238 (2000)
4. Kudo, T., Yamamoto, K., Matsumoto, Y.: Applying Conditional Random Fields to Japanese Morphological Analysis. In: Proceedings of Conference on Empirical Methods in Natural Language Processing 2004, pp. 230–237 (2004)
5. Deerwester, S., Dumais, S.T., Furnas, G.W., Landauer, T.K., Harshman, R.: Indexing by Latent Semantic Analysis. Journal of the Society for Information Science 41, 391–407 (1990)

6. Niu, Z., Shi, S., Sun, J., He, X.: A Survey of Outlier Detection Methodologies and Their Applications. In: Deng, H., Miao, D., Lei, J., Wang, F.L. (eds.) AICI 2011, Part I. LNCS, vol. 7002, pp. 380–387. Springer, Heidelberg (2011)
7. Yang, J., Zhong, N., Yao, Y., Wang, J.: Peculiarity Analysis for Classifications. In: Proceeding of IEEE International Conference on Data Mining, pp. 608–616 (2009)
8. Buchanan, M.: Ubiquity: Why Catastrophes Happen. Broadway (2002)

System and Its Method for Reusing an Application

Kwang-Yong Kim, Il-Gu Jung, and Won Ryu

Intelligent Convergence Media Research Department, ETRI 218 Gajeongno, Yuseong-gu, Daejeon, Korea
{kwangyk,ilkoo,wlyu}@etri.re.kr

Abstract. In the open USN(Ubiquitous Sensor Network) service platform, USN application only requests to the open USN service platform and remaining processing is done by the open USN service platform. But, the Open USN service platform which being developing as the international standards needs to consider an open interface for a service request of different service providers. Therefore, the Open USN service platform for heterogeneous applications by combining existing applications require interfaces that can be reused. We designed the system which could infer and recombine the application which it can reuse from the pre-registered applications. Thus, we can reuse pre-registered applications so that can reduce the cost and effort for implementing a duplicated application. Besides, there can be an advantage which has the function of one source multi use.

1 Introduction

Today, there are many available middleware for sensor networks and different kinds of Ubiquitous Sensor Network(USN) middleware may be deployed. Besides, USN services may utilize widely distributed sensors or sensor networks through different USN middleware. In this widely distributed environment, USN applications need to know of the various USN middleware, sensors and sensor networks used. The open USN service platform aims to provide unified access for heterogeneous USN middleware, thereby enabling USN applications to take full advantage of the USN capabilities. It would allow providers, users or application developers to provide USN resources, use USN services or develop USN applications without needing to have specific knowledge about the USN middleware and sensors or sensor networks of access. The three purposes of open USN service platform is to provide easy access to and use of the global USN resources and sensed data, easy connection of USN resources and easy development and distribution of various applications in last. Thus, in the open USN service platform, each application does not need to know how to access heterogeneous USN middleware nor which USN resources should be accessed. But there is a problem in this open USN service platform. It is required to provide open interface for heterogeneous services and applications to be used a new service or application through the combination of existing services and applications. This function reduces the necessary cost for the implementation of a new application and service by using existing services. Now we propose the pre-registered application

K.M. Lee, S.-J. Park, J.-H. Lee (eds.), *Soft Computing in Big Data Processing*, Advances in Intelligent Systems and Computing 271,
DOI: 10.1007/978-3-319-05527-5_2,

based reuse system for supporting the heterogeneous application. The remainder of the paper is organized as follows: Section II describes the background of study for the open USN service platform and describes requirements and the functional architecture for the open USN service platform which are recommended in ITU (International Telecommunication Union)-T[1]. Where, the Study Groups of ITU's Telecommunication Standardization Sector (ITU-T) are consist of experts from around the world to develop the international standards known as ITU-T Recommendations which act as defining elements in the global infrastructure of information and communication technologies[3]. Section III describes the design of the pre-registered application based reuse system for supporting the heterogeneous application and finally we provide some conclusions in the Section VI.

2 The Trends of Related Research

2.1 Introduction to the Open USN Service Platform

USN services require USN applications to have knowledge of USN middleware and sensors or sensor networks in order to access USN resources[1][2]. For example, heterogeneous USN middleware are not easily accessed by applications since each USN middleware may have proprietary Application Programming Interfaces (APIs) and may hinder access to various USN resources attached to the USN middleware. Even when the applications have access to multiple USN middleware, the applications have to search, collect, analyze and process the sensed data by themselves[1][2]. In the USN service framework, each application needs to know how to access heterogeneous USN middleware and which USN resources should be accessed[2]. In the open USN service framework, each application does not need to know how to access heterogeneous USN middleware nor which USN resources should be accessed[1]. Thus, In the open USN service platform, USN application only requests to the open USN service platform and remaining processing is done by the open USN service platform. The open USN service platform converts a request from application into the specific request for different USN middleware[1]. The following are the Requirements to communicate with heterogeneous USN middleware[1]. First, It is required to provide open interface for heterogeneous USN middleware to provide the sensed data and metadata received from USN resources. Secondly, USN resources and sensed data are required to be identified by Universal Resource Identifier (URI). Thirdly, URI for USN resource is required to be dynamically assigned when the USN resource is registered to the open USN service platform. Finally, it is required to provide standardized interface for accessing heterogeneous USN middleware. Figure 1 shows multiple heterogeneous USN middleware access provided by the open USN service platform[1]. There are also three requirements and two recommendations for the open USN service platform. In first, the requirements for the open USN service platform is as follows. Firstly, USN resource management is required according to proper management policies on authentication, authorization and access rights. Secondly, the characteristics and status of USN resources are required to be managed. Thirdly, it is required to provide functionality of inheritance and binding of USN middleware management policy. Besides, there is also recommendations that it may manage a logical group of USN resources according to service requests and may provide inference functions to derive the context data by users' rules.

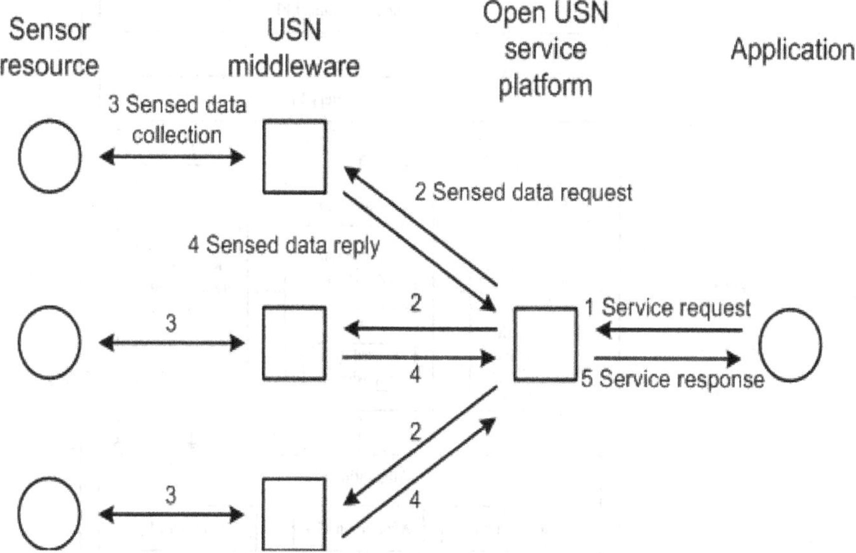

Fig. 1. Heterogeneous USN middleware access through the open USN service platform

2.2 Functional Architecture of the Open USN Service Framework

When we look at the a international standard draft which is recommended in ITU-T F.OpenUSN, The functional architecture of the open USN service framework largely consists of the open USN service platform and heterogeneous USN middleware.[1] The former (that is, open USN service platform) consists of seven functional entities (FEs): Application support FE, LOD(Linking Open Data) linking FE, Semantic inference FE, Resource management FE, Semantic USN repository FE, Semantic query FE and Adaptation FE. The latter(that is, The heterogeneous USN middleware) are integrated into the open USN service platform through the Adaptation FEs, furthermore the metadata of USN resources and semantic data are shared with the other services through LOD linking FE. Figure 2 shows functional architecture of the open USN service framework which is recommended in ITU-T F.OpenUSN. The functional role of each entity in the open USN service platform are as follows :

- Application support FE: It provides the functions which enables USN application to obtain open USN services and/or the sensed data/semantic data from the open USN service platform. it also supports the functions which allow the establishment or maintenance of connections or disconnection according to the type of a request for data, and access control to handle access rights for user authentication and the use of services.

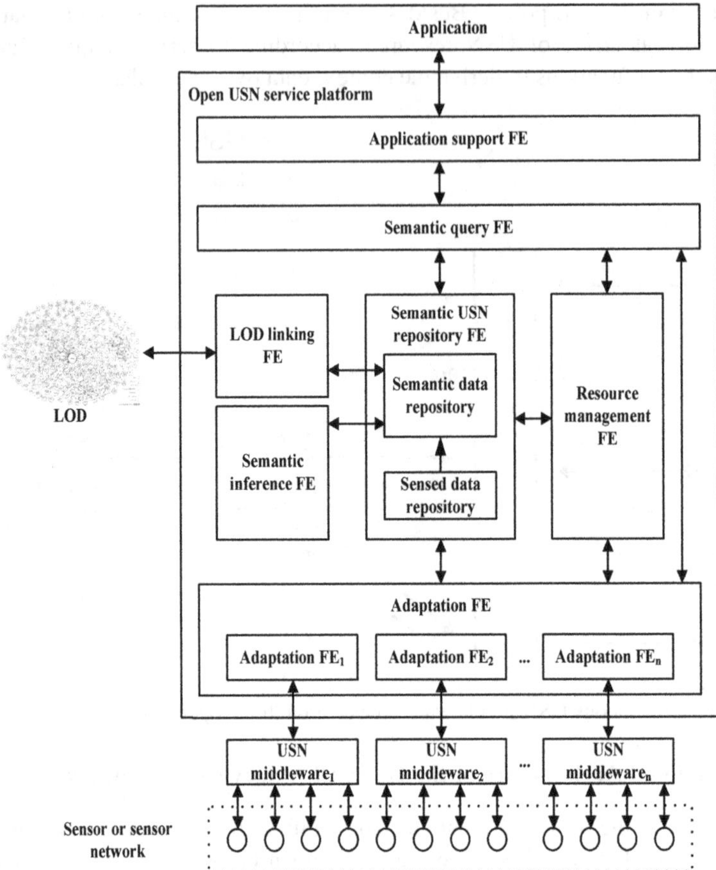

Fig. 2. Functional architecture of the open USN service framework

- LOD linking FE : It performs the functions which enable users to access the metadata of USN resources and semantic data from the open USN service platform to the LOD via Web, and which allow to link external LOD data with the metadata of USN resources and semantic data from the open USN service platform. It also supports the interface for query about the metadata of USN resources and semantic data in the LOD, and the functions which allow the application and management of policies that include criteria about selection and publication of data for the LOD.
- Semantic inference FE (optional): It provides the inference functions based on the information described in the ontology schema and users' rules by using the data stored in the Semantic USN repository FE. It also performs the functions to compose different kinds of patterns and levels for inference.
- Resource management FE : It provides the functions to issue and manage the URI of USN resource and sensed data; it also provides the functions to manage

the mapping relations with the address of USN resource. It further supports the functions which enable USN resource to be automatically registered in the open USN service platform when USN resource is connected in the network such as Internet, and applications to obtain and utilize information about USN resource. It provides the functions which enable USN resource to actively register its own status and connection information. It provides the functions to search URIs of USN resources for performing queries which can provide necessary information for requests from applications. Also, it supports the functions to create, maintain and manage information such as the purpose of the resource group, makers, control with right and so on. It provides the functions to manage the lifecycle of each resource group according to the duration of service.

- Semantic USN repository FE : It provides functions of converting metadata of USN resources and sensed data into RDF form. Semantic USN repository FE includes two different repositories called Semantic data repository and Sensed data repository. Furthermore, it stores sensed data collected from USN middleware to the Sensed data repository. It also provides functions to query for inserting new data, searching, modifying and deleting stored data.
- Semantic query FE : It performs the functions that handle queries to USN middleware and Semantic USN repository FE for providing responses to applications' information requests.
- Adaptation FE : It provides the functions which handle the protocol and message for setting up connection with USN middleware and delivering queries and commands. It works as an interface for processing several types of data generated from heterogeneous USN middleware in the open USN service platform. It also supports the message translation function to translate the generated data from heterogeneous USN middleware according to proper message specifications to deal with in the open USN service platform.

3 The Design of the Pre-registered Application Based Reuse System[4]

As we discussed in the previous chapter, In the open USN service platform, each application does not need to know how to access heterogeneous USN middleware nor which USN resources should be accessed. But there is a problem in this open USN service platform. It is required to provide open interface to services and applications for the combination of existing applications. Figure 3 shows the proposed functional block diagram which we can support the Pre-registered Application Based Reuse System. An advantage of the Pre-registered Application Based Reuse System is that there can be decrease the cost and effort for the implementation of a new application and has the function of one source multi use. Now, we explain the processing procedure for the Pre-registered Application Based Reuse System according to the each step.

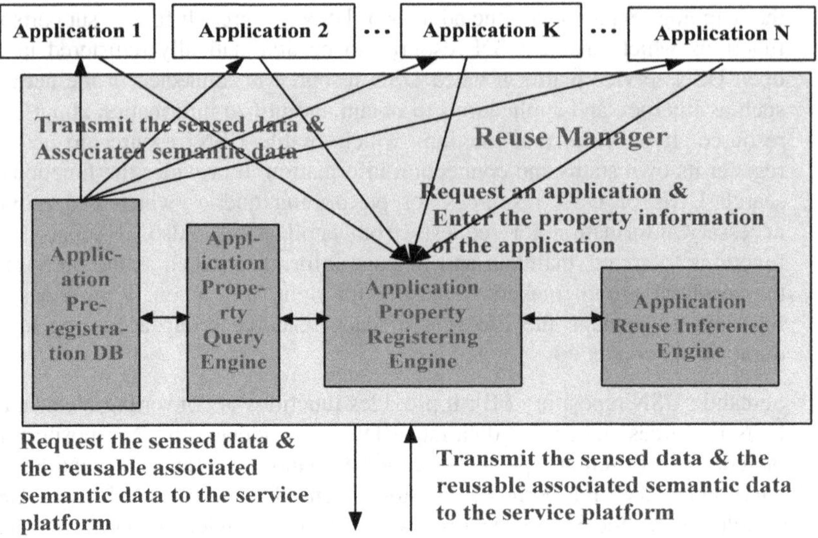

Fig. 3. The functional block diagram of the pre-registered application based reuse system

3.1 The Processing Procedure of the Pre-Registered Application Based Reuse System

Figure 4 shows the processing flow chart of the Pre-registered Application Based Reuse System, which we have proposed.

- 1st Step: The Application reqeuster enters the property information using the Application Property Registering Engine of the Reuse Manager and creates the standardized application profile for requests of the heterogeneous application. Where, Each application requesters who requesting the applications enter the application property information which requested by the Application Property Registering Engine using the property information window which the Reuse manager provides at the first registration. The Reuse manager creates the application profile which recorded the property information. The application profile includes the various properties information which can determine whether the it is reusable application or not. For example, it includes the various property information such as the service type, the service provider's name and service properties.
- 2nd Step: The Application Property Query Engine retrieves the application property information of the packaging data which stored in the Application Pre-registration DB by using the application query tool. A package data consists of the sensed data and of the semantic data and of the application's profile. This package data is stored as the structure of the ontology based layered tree and is managed by the Reuse manager.

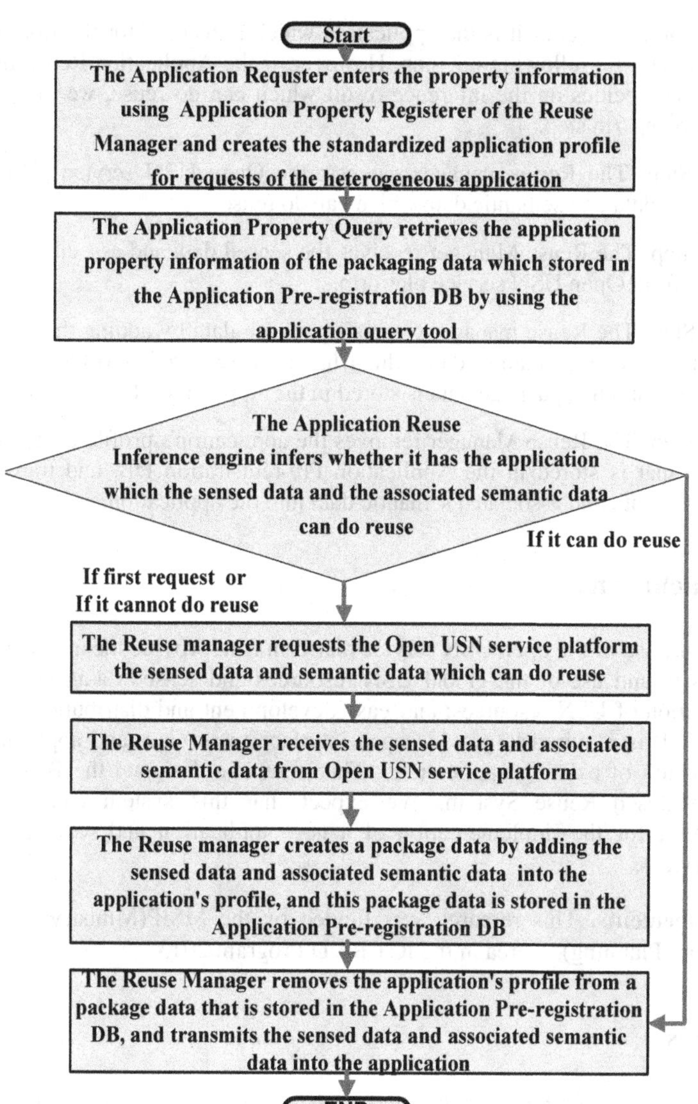

Fig. 4. The processing flow chart for the pre-registered application based reuse system

- 3rd Step: The Application Reuse Inference engine compares the application's profile produced in 2nd step with the application's profile of a package data stored in the Application Registration DB. and it infers whether it has the application which the sensed data and the associated semantic data can do reuse. Where, the inference rules and the knowledge bases are contained in the Application Reuse Inference Engine to determine if the requested application can use by recombining the existing applications. In 3rd step, if the Application Reuse Inference Engine decides on the inference result which cannot do reuse, or it is the application which registered for the first time, we moves to the following 4th step. However, if the Application Reuse Inference Engine decides on the inference result which can do reuse, we moves to the following 7th step.

- 4th Step: The Reuse manager requests the Open USN service platform the sensed data and semantic data which can do reuse.
- 5th Step: The Reuse Manager receives the sensed data and associated semantic data from Open USN service platform.
- 6th Step: The Reuse manager creates a package data by adding the sensed data and associated semantic data which has been received into the application's profile, and this package data is stored in the Application Pre-registration DB.
- 7th Step: The Reuse Manager removes the application's profile from a package data that is stored in the Application Pre-registration DB, and transmits the sensed data and associated semantic data into the application.

4 Conclusions

The main purpose of the open USN service platform is to provide the application with easy access to and use of the global USN resources and sensed data/semantic data; easy connection of USN resources; and easy development and distribution of various applications. It is required to provide open interface to services and applications for the combination of existing applications. Therefore, we designed the Pre-registered Application Based Reuse System. We expect that this system can reduce the necessary cost for the implementation of a new application and service by using existing services.

Acknowledgements. This research was funded by the MSIP(Ministry of Science, ICT & Future Planning), Korea in the ICT R&D Program 2013

References

1. Lee, J.S., et al.: F.OpenUSN Requirements and functional architecture for open USN service platforms (New): Output draft, ITU-T (WP2/16), TD32 (2013)
2. Lee, H.K., et al.: Encapsulation of Semiconductor Gas Sensors with Gas Barrier Films for USN Application. ETRI Journal 34(5), 713–718 (2012)
3. http://www.itu.int/en/ITU-T/about/Pages/default.aspx
4. Kim, K.Y., et al.: F.OpenUSN: Proposed updates, ITU-T (IoT-GSI), TD190 (2013)

Assessment of Scientific and Innovative Projects: Mathematical Methods and Models for Management Decisions

Galim Mutanov[1], Gizat Abdykerova[2], and Zhanna Yessengaliyeva[1]

[1]Department of Mechanics and Mathematics, al-Farabi Kazakh national university 71, al-Farabi ave., Almaty, 050040, Kazakhstan
Rector@kaznu.kz, Zhanna_Serzhan_7@mail.ru
[2]Economics Department, D. Serikbayev East Kazakhstan state technical university 69, Protozanov str., Ust-Kamenogorsk, 070004, Kazakhstan
kanc_ekstu@mail.ru

Abstract. To improve the quality of assessment of scientific and innovative projects the mathematical methods and models are developed in this article. Criteria and methods for assessing innovativeness and competitiveness are developed, along with a graphic model allowing visualization of project assessment in the coordinate scale of the matrix model. Innovation projects being objects of two interacting segments – science and business – are formalized as two-dimensional objects with the dependence K=f(I), where K is competitiveness and I is innovativeness. The proposed methodology facilitates complex project assessment on the basis of absolute positioning. This decision support methodology is designed to be utilized by expert commissions responsible for venture funds, development institutes, and other potential investors needing to select appropriate scientific and innovative projects.

Keywords: scientific and innovative project, innovation management, ranking, competitiveness, innovativeness

1 Introduction

The process of introducing a novelty product into the market is one of the key steps of any innovative process, which results in an implemented/realized change - innovation. Scientific and innovative project (SIP) as a subject of analysis and appraisal. SIP are long-term investment projects characterized by a high degree of uncertainty as to their future outcomes and by the need to commit significant material and financial resources in the course of their implementation [1].

The life cycle of a scientific and innovative project can be defined as the period of time between the conception of an idea, which is followed by its dissemination and assimilation, and up to its practical implementation [2].

K.M. Lee, S.-J. Park, J.-H. Lee (eds.), *Soft Computing in Big Data Processing*, Advances in Intelligent Systems and Computing 271,
DOI: 10.1007/978-3-319-05527-5_3, © Springer International Publishing Switzerland 2014

Another way of putting it is that the life cycle of an innovation is a certain period of time during which the innovation is viable and brings income or other tangible benefit to a manufacturer/seller of the innovation product.

The specific features of every stage of the innovation life cycle.

Innovation onset. This stage is important in the entire life cycle of a product. Creation of an innovation is a whole complex of works related to converting the results of scientific research into new products, their assimilation in the market, and involvement in trade transactions. The initial stage of a SIP is R&D, during which it is necessary to appraise the likelihood for the project to reach the desired performance.

Implementation. This stage starts when a product is brought into the market. The requirement is to project the new commodity prices and production volume.

Growth stage. When an innovation fits market demand, the sales grow significantly, the expenses are reimbursed, and the new product brings profit. To make this stage as lengthy as possible, the following strategy should be chosen: improve the quality of an innovation, find new segments in the market, advertise new purchasing incentives, while bringing down the prices.

Maturity stage. This stage is usually the most long-lived one and can be represented by the following phases: growth retardation – plateau – decline in demand.

Decline stage. At this stage, we see lower sales volumes and lower effectiveness.

Given the above sequence, an innovation life cycle is regarded as the innovation process.

An innovation process [3] can be considered as the process of funding and investing in the development of a new product or service. In this case, it will be a SIP as a particular case of investment projects, which are widely used in the business environment. Understanding fundamentals and life cycle of innovation process is decisive for developing appropriate frameworks that aim to improve success of innovation process [4].

Thus, the life cycle concept plays a fundamental role in planning the innovative product manufacturing process, in setting business arrangements for an innovation process, as well as during project design and evaluation stages.

2 Materials and Methods

Appraisal of a SIP is an important and challenging procedure at the research and development stage [5]. It is a continuous process that implies a possible suspension or termination of a project at any point of time when new information is obtained.

In this section, we present mathematical methods and models of an assessment of scientific and innovative projects, whether a SIP is feasible or not, the project must go through several stages of examination, which are shown in Figure 1.

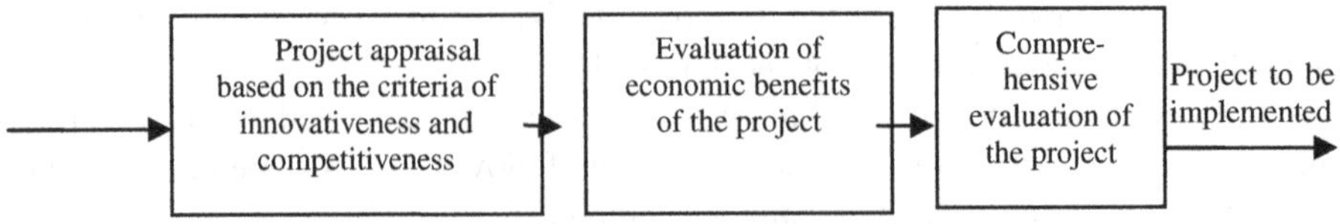

Fig. 1. Scientific and innovative project appraisal stages

First stage of appraisal is a method of assessment of SIP referred to the scientific, technical, and industrial sector, with a system of target indicators. The relationship between competitiveness and innovativeness originates from definitions of these concepts [6]. Competitiveness can be understood as the ability of companies to produce goods and services that can compete successfully on the world market [7,8]. In turn, innovativeness can be understood as introduction of a new or significantly improved idea, product, service designed to produce a useful result. The quality of innovation is determined by the effect of its commercialization, the level of which can be determined by assessing the competitiveness of products.

There is a certain relationship between competitiveness and innovativeness. SIP being objects of two interacting segments – science and business – are formalized as two-dimensional objects with the dependence like formula (1).

$$K=f(I), \tag{1}$$

where K is competitiveness and I is innovativeness. In a certain sense, innovative relations are the result of competitiveness, which enables us to consider competitiveness as a function of innovation. Innovativeness and competitiveness are the most meaningful indicators for the SIP appraisal process. To evaluate a SIP at the R&D stage, the following basic innovativeness criteria are suggested in Table 1:

Table 1. Criteria of innovativeness

	Innovativeness criteria
1	Compliance of a project with the priority areas of industrial and innovation strategy.
2	Relevance of research and product uniqueness (no analogues).
3	Scientific originality of the solutions proposed within the project.
4	Technological level of the project (technology transfer, new technology).
5	Advantages of the project in comparison with analogues existing in the world.
6	Economic feasibility of the project.

Based on the data available on competitiveness components, the competitiveness criteria can be schematically presented as follows indicators in the Table 2.

Table 2. Criteria of competitiveness

	Competitiveness criteria
1	Availability of markets and opportunities to commercialize the proposed results
2	Level of competitive advantages of R&D results to retain them in the long-run
3	Consistency with the existing sale outlets (distribution channels)
4	Patentability (possibility to defend the project by using the patent)
5	Availability of proprietary articles
6	Availability of scientific and technical potential of the project
7	Technical feasibility of the project
8	Project costs
9	Degree of project readiness
10	Availability of a team and experience in project implementation
11	Opportunities to involve private capital (investment attractiveness)
12	Scientific and technical level of project

Such set of criteria makes it possible to conduct an initial assessment of SIP. In developing the method we used a methodological approach based on expert assessment of innovation and competitiveness indicators for SIPs, accompanied by a graphic model of project innovativeness (I) and competitiveness (K) assessment.

Adequacy of the criteria for the complex index is determined by assigning weights to each criterion and using an additive–multiplicative method of calculation.

SIP assessment, based on the graphic model for assessing project I and K, should be carried out in three stages: a) selecting optimal criteria, b) determining weight coefficients, and c) positioning projects in the McKinsey matrix [9].

To calculate these criteria, we propose the following method. The way to solve this task is related to determination of the average expert values for each I and K criterion. Common values of criteria are defined as follows:

$$I_j = \sum_{i=1}^{n} x_i f_{ij}, \sum_{i=1}^{n} x_i = 1 \tag{2}$$

$$K_j = \sum_{k=1}^{m} y_k g_{kj}, \sum_{k=1}^{m} y_k = 1 \tag{3}$$

$$I_{min} \leq I_j \leq I_{max}, K_{min} \leq K_j \leq K_{max}, \tag{4}$$

where f_{ij} - is the value of i-th criterion of the j-th project for the innovativeness;

x_i– value of weighting coefficient of i-th criterion for the innovativeness indicator;

n– number of criteria for the innovativeness indicator;

g_{kj} – value of the k-th criterion of j-th object (project) for the competitiveness;

y_k– value of weighting coefficient of the k-th factor for the competitiveness indicator;

m– number of criteria for the competitiveness indicator;

$j=1,J$ – with J being the number of objects (projects);

I_{min}, I_{max}, K_{min}, K_{max} – minimum and maximum values of indicators.

In the graphic model for assessing project innovativeness and competitiveness, the range of indicators is split into 9 sectors. In this case, it is required to define I and K parameters, which are the coordinates of these projects in the matrix. To determine the coordinates in the model we use weighted average factors. It is recommended that the values for each factor be assessed using the expert approach (from 1 to 9); in the presence of several experts, the values are averaged.

To formalize the criteria rating, expert estimates are the most suitable tool. To determine the weighting coefficients for each criterion we used the ranking method.

While ranking, the initial ranks are transformed as following: each rank equal to quantity of criteria dividing to initial rank. Then totals are calculated by these transformed ranks:

$$R_J = \sum_{K=1}^{M} R_{JK} \div M, \tag{5}$$

where R_J - the average of the ranks converted across all the experts for j-th factor; R_{JK} - converted rank assigned by k-th expert to j-th factor; M - number of experts.

Next, the weights of criteria are calculated by formula (6):

$$W_J = R_J \div \sum_{J=1}^{N} R_J, \tag{6}$$

where W_J –average weight of criterion across all experts; N –number of criteria.

Consistency of expert assessments by criteria was verified by calculating the coefficients of factor variations, which are the analogous to dispersion and represented in the following formula (7):

$$S = \sum_{i=1}^{n}\left\{\sum_{j=1}^{m} x_{ij} - \frac{1}{2}m(n+1)\right\}^2, \tag{7}$$

where S - factor variation coefficient; x_{ij} - the rank of i-factor assigned by j-expert; m - number of experts; n - number of criteria.

Since experts come from various entities or groups, there is a need to identify homogeneity of these groups. To solve this problem across various criteria, derived from experts, the consistency of their views is determined by using concordance coefficient. Concordance coefficient (W) is calculated by using the formula (8) proposed by Kendall:

$$W = \frac{12 \cdot S}{m^2 \cdot (n^3 - n)}, \tag{8}$$

where S - the sum of squared differences (deviations); m - number of experts; n - number of criteria.

In the case where any expert fails to identify the rating value between several related factors and assigns the same rank, the concordance coefficient is calculated by using the following formula (9) and (10):

$$W = \frac{S}{\frac{1}{12}m^2(n^3 - n) - m\sum_{j=1}^{m} T_j}, \tag{9}$$

$$T_j = \frac{1}{12}\sum_{t_j}(t_j^3 - t_j), \tag{10}$$

where tj – is the number of equal ranks in the series j.

In this way, this method allows prioritizing projects on such key indicators as innovativeness and competitiveness.

The second stage of project assessment is to determine the economic viability of the project, the methodology of which is described in the work [10].

The method of an assessment of efficiency of projects is based on following methods: Net present value (NPV) method of project evaluation; Profitability index (PI) method to estimate investment profitability; Internal rate of return (IRR); and Payback period (PB).

A more comprehensive method should be used at the **final project appraisal stage**. Having applied the two above methods, it is considered practical to appraise and compare alternative projects with a more comprehensive technique, which is based on determining vector lengths in the Cartesian system.

The 3D Cartesian coordinate system has three mutually perpendicular axes Ox, Oy, and Oz with the common zero point O and preset scale. Let V be an arbitrary point and Vx, Vy, and Vz be its Ox, Oy, and Oz components (Fig. 2). The coordinates of points Vx, Vy, Vz will be denoted as xV, yV, zV, respectively; that is, Vx(xV), Vy(yV), and Vz(zV). Thus, the point V has the following coordinates: V = (xV; yV; zV).

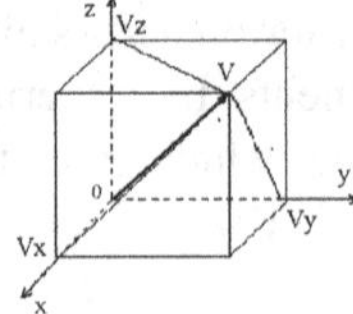

Fig. 2. Three-dimensional Cartesian system of coordinates

Then the length of vector OV will be calculated by the formula (11):

$$|V| = \sqrt{x^2 + y^2 + z^2}\,. \tag{11}$$

Thus, the comprehensive method can be used to appraise a SIP by determining the vector length in a 3D coordinate system (x, y, z), where x is the project innovativeness (I), y is the project competitiveness (K), and z is the net present value (NPV). The values I, K, and NPV are specified both in the innovativeness and competitiveness assessment method and in the economic benefit evaluation method.

The scales for I and K axes have been normalized to conventional standard units with due account for their commensurability and comparability. The scales for I and K axes are given as dimensionless values with a 9-point scoring system making it possible to classify projects into 3 groups, namely: outsiders, marginal, and leaders. This approach allows us to classify projects into groups based on their profitability and move to the dimensionless unit of measurements. Monetary units have also been expressed in dimensionless units by assuming that a 10 billion dollar project has its NPV equal to 3 conventional units, the NPV of a 20 billion dollar project will be 6 units, and that of a 30 billion dollar project will equal 9 units. If and when needed, this proportion may be changed as a function of resulting NPV values.

3 Experimental Results

The proposed methods were developed based on two scientific and innovative projects: Project #1, and Project #2. Let us consider how the proposed methods can be applied. The **first step** in SIP evaluation is project examination using the innovativeness and competitiveness criteria. Based on the above, to calculate criteria values the questioning of 22 experts was undertaken. In qualitative terms, the experts were managers, specialists of R&D, and innovation managers. The questionnaire was compiled based on the two sets of criteria outlined above. The total number of questionnaires was 22. This quantity is optimum as at very high number of experts there is a probability of loss of objectivity. Criteria ranking was quite simple: By assigning the highest ranking to the criterion of the lowest value, in their opinion. Importance (rank) of each criterion is determined by the average estimated value and the sum of ranks of expert assessments. The expert evaluation results make it possible to obtain weighting coefficients to determine positioning of SIP in the matrix. Weights are demonstrating the importance of each criterion. These data indicate that expert evaluations for each group of the criteria differ by their significance. For I indicators these weights are calculated, and presented in the Table 3:

Table 3. Calculation of weight coefficients and definition of ranks of estimation of SIP

	Criteria	Average of ranks	Weights	Rank
1	Compliance of a project with the priority areas of industrial and innovation strategy.	3,35	0,228	3
2	Relevance of research and product uniqueness.	3,70	0,252	1
3	Scientific originality of the solutions proposed with the project.	1,50	0,102	5
4	Technological level of the project.	1,54	0,105	4
5	Advantages of the project in comparison with analogues existing in the world.	1,02	0,069	6
6	Economic feasibility of the project.	3,59	0,244	2

For K indicators, weight coefficient values are as follows: 0.277, 0.119, 0.033, 0.060, 0.067, 0.067, 0.040, 0.037, 0.041, 0.142, 0.057, and 0.061. Within this group, the criteria are the following: the first one is availability of the market and opportunities to commercialize the proposed project results; the second, availability of a team of qualified specialists having experience in project implementation; the third, level of competitive advantages of R&D results and potential to retain them in the long-run, etc. W-Kendall concordance coefficient equal 0.69, that means good degree of coherence of rankings.

Further, involves positioning of projects within the graphic model of innovativeness and competitiveness of SIP. The assessments of I and K indicators averaged across five experts. Based on these averaged assessments and weighting coefficients, SIPs were positioned in the graphical model of SIPs. The obtained weights and assessment criteria on the innovativeness indicators are presented in Tables 4. Utility estimates for Projects 1 and 2 based on K indicators calculated by analogy.

Table 4. Utility estimate for Projects 1 and 2 based on the innovativeness indicators

Criteria of innovativeness	weights	Criteria values (average) Project #1	Criteria values (average) Project #2	Normalized estimate of priority vector, Project #1	Normalized estimate of priority vector, Project #2
1	0.228	3.6	4.2	1.56	1.96
2	0.252	3.6	3.6	1.27	2.12
3	0.102	4	3.8	0.44	0.90
4	0.105	3.8	3.2	0.38	0.88
5	0.069	3	3.2	0.22	0.57
6	0.244	2	3.8	1.67	2.10
Total	**1**	**20**	**21.8**	**5.54**	**8.52**

Figure 3 demonstrates graphic model for assessment of project innovativeness and competitiveness. The proposed projects were positioned according to expert assessments received.

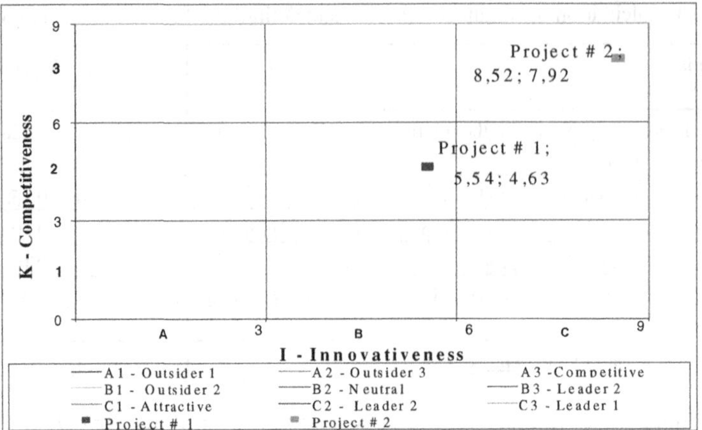

Fig. 3. Example of projects positioning in graphic model

The resulting matrix allows positioning each SIP based on the criteria of I and K indicators in a certain sector. Matrix boundaries are the maximum and minimum possible values from 1 to 9, respectively. Three groups are highlighted in this matrix: leader, outsider, and border. Projects that fall into group of "leaders" have the highest values of I and K indicators as compared with the other two groups; they are the absolute priority to be implemented at the earliest possible time. Projects that fall into the three sections in the lower left corner of the matrix "outsiders" have low values based on many criteria. These projects are problematic. The three sections located along the main diagonal, going from lower left to upper right edge of the matrix have the classical name of "border": these include the competitive sector and neutral. These projects are requiring finalization work. It follows from the obtained positioning that, judging by the normalized priority vectors, Project #2 has been classified in the "leader" group, which means that it is a priority project and can be implemented. Project #1 is in the "neutral" group meaning that it should be further engineered and finalized. Project #2 moves to the next stage of estimation.

The second step of the project examination procedure is the economic effect analysis of a project. The project NPV for the total life of the project is 12,506,094,589.24. The results of the economic effect are summarized in the Table 5.

Table 5. Main indicators of the project's economic effect

Indicator	USD	Indicator	USD
Discount rate, %	11.29	NPV	12,506,094,589.24
Payback period (PB), months	22	Profitability index (PI)	21.42
Discounted payback period (DPB),	25	IRR (%)	320.25

The calculations show that the discounted payback period is 25 months, which is quite plausible for such projects. The profitability index (PI) is much higher than 1.0.

The IRR is also high. Therefore, the project has sufficiently high indices of effectiveness and can be accepted for implementation.

The third step of examining a SIP is the proposed comprehensive method. The I, K, and NPV values are 6.73, 6.84, and 12.5, respectively. The NPV value of USD 12.5 billion as expressed in conventional standard units will be 3.75. Now we go on to defining the vector length for Project #2 presented in Figure 4.

$$|V| = \sqrt{I^2 + K^2 + NPV^2} = \sqrt{6.73^2 + 6.84^2 + 3.75^2} = 10.3 \cdot$$

Therefore, the indices of effectiveness of Project #2 are sufficiently high and it can be accepted for implementation. This method works effectively when ranking alternative projects. The longer the vector, the more chances for the project to be accepted; that is, the absolute positioning approach is used. It follows from the above that all three methods used for examining the project show that it is acceptable and can be implemented. In the above example, Project #2 has high estimated performance indicators and can be accepted for implementation.

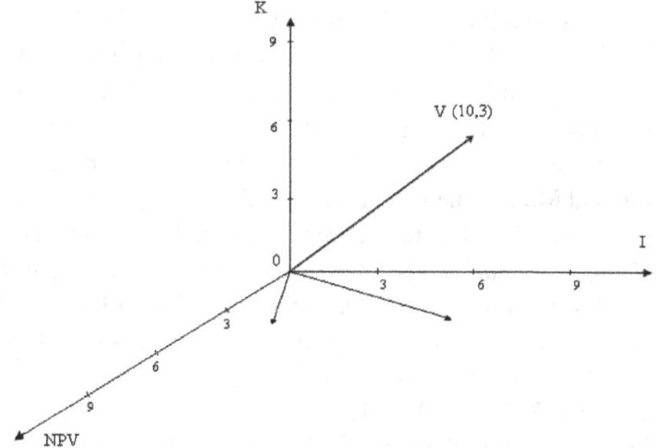

Fig. 4. Graphic model of the comprehensive evaluation of innovation projects

4 Discussion and Conclusion

To achieve competitive advantage, firms must find new ways to compete in their niche to enter the market in the only possible way: through innovation. Thus innovation equally includes R&D results of production purposes, and results geared at improving the organizational structure of production [3]. Our results suggest that R&D efforts play an important role [11] in affecting product innovation. As a solution, these methods and models let us select the project whose normalized estimates of priority vectors by value occupy the “Leader 1” section.

In order to protect the safety of financial investments in conditions of information uncertainty, methods and mathematical models are developed to assess alternative SIPs. A richly complex method of assessment of SIP based on such parameters as I, K, and NPV of a project facilitate in-depth assessment of SIPs on the basis of absolute positioning. The methods and models presented here can be used by experts of venture funds, institutes, and other potential investors to meaningfully assess SIPs.

Acknowledgements. We gratefully acknowledge the financial support of the Committee of Science of The Ministry of Education and Science of Republic of Kazakhstan which enabled this study. Grant number 1150 / GF 2012-2014.

References

1. Hochstrasser, B., Griffths, C.: Controlling IT Investment: Strategy and Management. Chapman & Hall, London (1991)
2. Powell, W., Grodal, S.: Networks of Innovators. In: The Oxford Handbook of Innovation, vol. 29, pp. 56–85. Oxford University Press, Oxford (2005)
3. Christensen, C.: The Innovator's Dilemma. Harvard Business School Press (1997)
4. Dereli, T., Altun, K.: A novel approach for assessment of technologies with respect to their innovation potentials. Expert Systems with Applications 40(3), 881–891 (2013)
5. Takalo, T., Tanayama, T., Toivanen, O.: Estimating the benefits of targeted R&D subsidies. Review of Economics and Statistics 95(1), 255–272 (2013)
6. van Hemert, P., Nijkamp, P.: Knowledge investments, business R&D and innovativeness of countries. Technological Forecasting and Social Change 77(3), 369–384 (2010)
7. Sebora, T., Theerapatvong, T.: Corporate entrepreneurship: A test of external and internal influences on managers' idea generation, risk taking, and proactiveness. International Entrepreneurship and Management Journal 6(3), 331–350 (2010)
8. Porter, M.: The competitive advantage of nations. Free Press, New York (1990)
9. Kurttila, M.: Non-industrial private forest owners' attitudes towards the operational environment of forestry. Forest Policy and Economics 2(1), 13–28 (2001)
10. Mutanov, G., Yessengaliyeva, Z.: Development of methods and models for assessing feasibility and cost-effectiveness of innovative projects. In: Proc. of Conf. SCIS-ISIS 2012, Kobe, Japan, pp. 2354–2357 (2012)
11. Yin, R.: Case Study Research. Sage Publications, Beverly Hills (1994)

Multi-Voxel Pattern Analysis of fMRI Based on Deep Learning Methods

Yutaka Hatakeyama[1,*], Shinichi Yoshida[2], Hiromi Kataoka[1], and Yoshiyasu Okuhara[1]

[1] Center of Medical Information Science, Kochi University, Japan
{hatake,kataokah,okuharay}@kochi-u.ac.jp
[2] School of Information, Kochi University of Technology, Japan
yoshida.shinichi@kochi-tech.ac.jp

Abstract. A decoding process for fMRI data is constructed based on Multi-Voxel Pattern Analysis (MVPA) using deep learning method for online training process. The constructed process with Deep Brief Network (DBN) extracts the feature for classification on each ROI of input fMRI data. The decoding experiment results for hand motion show that the decoding accuracy based on DBN is comparable to that with the conventional process with batch training and that the divided feature extraction in the first layer decreases computational time without loss of accuracy. The constructed process should be necessary for interactive decoding experiments for each subject.

Keywords: fMRI, MVPA, deep learning, Deep Brief Network.

1 Introduction

Decoding for fMRI data has been studied with the progress of analysis technology recently. Early studies of analysis for fMRI data mean coding process which finds the brain functional mapping based on relations between stimuli of task and activation in fMRI data. On the other hand, decoding methods [1 - 3] estimate tasks for stimuli or subject's states based on activation data in input fMRI data.

In decoding methods for fMRI data, Multi Voxel Pattern Analysis (MVPA) [3] is generally used, where several voxel values are used for classification. Preprocessing for MVPA is necessary for improvement of classification accuracy. That is, decoding process is divided into 3 processes, voxel selection, feature extraction, and classification. Selection and extraction processes are calculated before classification process and executed by some statistics, e.g. t-value, defined by all fMRI data in experiments. Therefore, it is difficult to apply these processes to the experiments which need interaction of each subject because the preprocessing cannot be done in each task.

In computer vision, deep learning methods [4, 5] have been proposed for image recognition, which the feature of input data is defined by not human-crafted but deep learning method. The deep learning methods consist of multi-layer artificial neural network. The training process is executed by unsupervised learning in each layer

* Corresponding author.

K.M. Lee, S.-J. Park, J.-H. Lee (eds.), *Soft Computing in Big Data Processing*, Advances in Intelligent Systems and Computing 271,
DOI: 10.1007/978-3-319-05527-5_4, © Springer International Publishing Switzerland 2014

which extracts feature values from output in the previous layer. From the results in these methods, the multi-layer methods can extract complex feature automatically.

This paper constructs decoding process based on Deep Brief Network (DBN), one of deep learning methods, in order to calculate feature for classification automatically. Considering an assumption that low level feature in each ROI of fMRI is not affected by other ROIs, the first layer of the DBN is processed in input values on every ROI regions. In order to check the validity of the constructed decoding process, the decoding experiments of hand motion are executed. The decoding accuracy is compared with that of the conventional process with batch preprocessing based on t-values. The computational time of the DBN method which has divided first layers is compared with that of the DBN method with one first layer.

The decoding process based on DBN in this paper is defined in 2. The decoding experiments of hand motion are shown in 3.

2 Decoding Process Based on DBN

Classification for fMRI data in decoding process is operated based on Deep Brief Network (DBN) for each ROI region in input voxels. Restricted Boltzmann Machine and DBN are introduced in 2.1 and 2.2, respectively. The decoding process for each ROI region is proposed in 2.3.

2.1 Restricted Boltzmann Machine

The Restricted Boltzmann Machine (RBM) is an artificial neural network constructed by bipartite graph with visible layer and hidden layer. That is, there exist symmetric connections between visible layer and hidden layer. Moreover, the RBM is a Boltzmann machine without connections in the same layers. A model of RBM is shown in Fig.1. In general, the hidden unit is defined by a binary unit. The visible unit for input data, e.g. voxel values, is defined by continuous unit in [0, 1]. In the other case, the visible unit is defined by a binary unit. The RBM is defined as a probabilistic generative model. The conjunctive probability of hidden unit values $\mathbf{h}$ and visible unit values $\mathbf{v}$ is defined by

$$P(\mathbf{v},\mathbf{h}) = \frac{1}{Z}\exp(-E(\mathbf{v},\mathbf{h})),$$

$$Z = \sum_{v}\sum_{h}\exp(-E(\mathbf{v},\mathbf{h})). \tag{1}$$

As shown in equation (1), the probability is defined using an energy function E. The normalizing factor Z is called the partition function by analogy with physical systems. The energy function in the binary visible units is defined by

$$E(\mathbf{v},\mathbf{h}) = -\sum_i b_i v_i - \sum_j c_j h_j - \sum_i \sum_j v_i W_{i,j} h_j \,. \tag{2}$$

In the continuous visible units, the function is defined by

$$E(\mathbf{v},\mathbf{h}) = \frac{1}{2}\sum_i v_i^2 - \sum_i b_i v_i - \sum_j c_j h_j - \sum_i \sum_j v_i W_{i,j} h_j \tag{3}$$

The parameters b_i, c_j, and $W_{i,j}$ mean the bias parameter of visible units, hidden units, and the weight parameter of the network, respectively. The RBM is processed as the generative model with unsupervised learning using visible unit values. The training process for RBM is operated by maximum likelihood estimation method for $P(\mathbf{v}) = \sum_h P(\mathbf{v},\mathbf{h})$. The conditional probability, $P(\mathbf{v} \mid \mathbf{h})$ and $P(\mathbf{h} \mid \mathbf{v})$, in which there is independent for each same unit, is calculated by

$$P(\mathbf{v} \mid \mathbf{h}) = \prod_i P(v_i \mid \mathbf{h}),$$

$$P(\mathbf{h} \mid \mathbf{v}) = \prod_j P(h_j \mid \mathbf{v}). \tag{4}$$

The probability is defined by

$$P(h_j = 1 \mid \mathbf{v}) = \sigma\left(c_j + \sum_i v_i W_{i,j} \right), \tag{5}$$

$$P(v_i = 1 \mid \mathbf{h}) = \sigma\left(b_i + \sum_j W_{i,j} h_j \right), \tag{6}$$

$$P(v_i \mid \mathbf{h}) = N\left(v_i ; b_i + \sum_j W_{i,j} h_j , 1 \right) \tag{7}$$

The probabilities for a binary visible unit and a continuous visible unit are shown in equation (6) and (7), respectively. A sigmoid function is denoted by σ. A probabilistic density of x in normal distribution with mean μ and variance 1 is denoted by $N(x;\ \mu, 1)$.

The training process is operated based on Stochastic Gradient Descent (SGD). Although it is difficult to calculate $P(v,h)$ compared with calculating $P(v \mid h)$ and $P(h \mid v)$, it is possible to obtain the practically approximate values using Contrastive Divergence (CD). After training, the hidden unit values are considered as feature extracted from visible unit values. Therefore, the classification for fMRI data is processed based on hidden unit values.

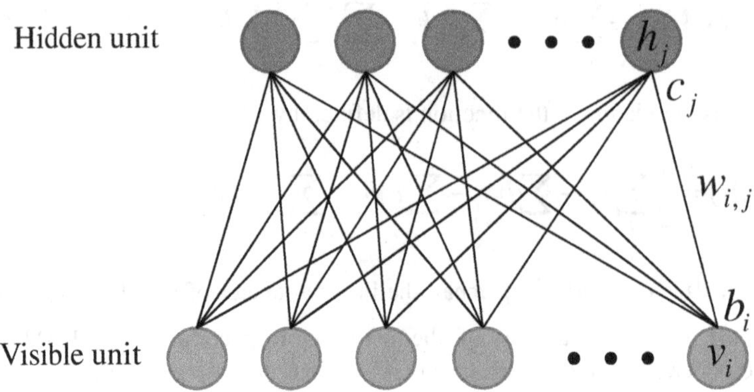

Fig. 1. A model of Restricted Boltzmann Machine

2.2 Deep Brief Network

As mentioned in 2.1, the hidden unit values are considered as features of visible unit values. However, the feature is considered as low level feature. If the classification process needs the complex feature values, the artificial neural network calculates the features based on the hidden values with multistage methods.

Deep Brief Network (DBN) is a multilayer artificial neural network constructed by several RBMs and a classifier. The sample model is shown in Fig.2. The first layer of DBN is deal with visible layer of the first RBM. The first layer and the second layer are trained by RBM training based on the input data for the DBN. Next, the parameters of the first RBM are fixed and the hidden unit values in second layer are used as visible unit values for the second RBM. The third layer of DBN is used as hidden layer of the second RBM. The DBN training process repeats these processes. That is, the DBN training consists of training with the each RBM. From these training processes, the DBN extracts feature from input values.

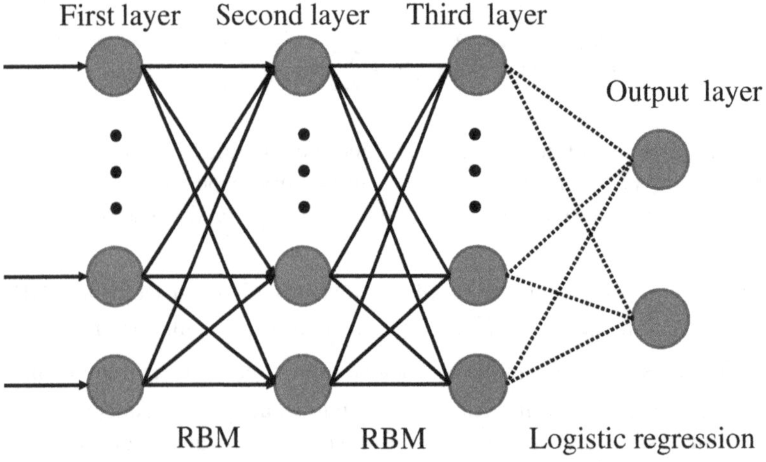

Fig. 2. A model of Deep Brief Network with logistic regression

In order to run the discriminative process for input, the classification process is used in the top layer of the DBN. The output layer is denoted by '1 of K' representation. In Fig.2, the number of classes for discrimination is 2, which the training of this layer is based on multinomial logistic regression. In general, the fine-tuning process for classification is operated by logistic regression in which the parameters of each RBM are possible to be modified. From these training processes, it is possible to construct the multi-layer artificial neural network for the input which has complex features.

2.3 Decoding Process for fMRI Data

The input values in general decoding processes are extracted from fMRI data in each ROI region which is considered as different function in brain. One of input styles for the DBN in decoding process is to input the values into a RBM in the DBN as shown in Fig.3. This style shows an assumption that all voxel values in each ROI are influenced by one another. This assumption should be valid from the viewpoints of the functional connectivity in brain. With consideration of multilayer process in the DBN, the multilayer manner means extracts a complex feature from input fMRI data. The model of one RBM style is shown in Fig.3.

As shown in equation (5)-(7), the calculation of visible unit values and hidden unit values is operated with double sum for each unit. Therefore, it takes more computational costs to train RBMs in the DBN with many units in each layer. That is, it is difficult to decode the input data with high speed. As above mentioned, lower layer

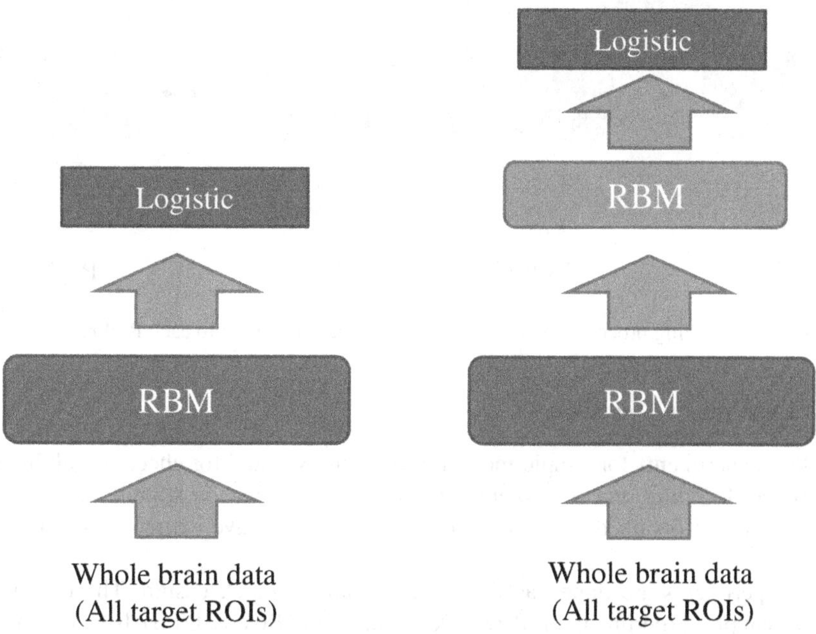

Fig. 3. Decoding process in which the first RBM uses all ROI data

RBM in the DBN is considered to extract the low level feature of the input data. This paper assumes that lower level feature of each ROI in fMRI data is not affected by the input values in different ROI. Based on this assumption, lower level layer in the DBN is divided into several ROI data. This model with two ROI regions is shown in Fig.4. Based on the hidden unit values in lower layer, the DBN integrates these features in each ROI data.

The classification process in this paper is processed by logistic regression in order to decode the subject's tasks.

3 Decoding Experiments

Decoding experiments of hand motions is executed to check the validity of decoding process based on the DBN.

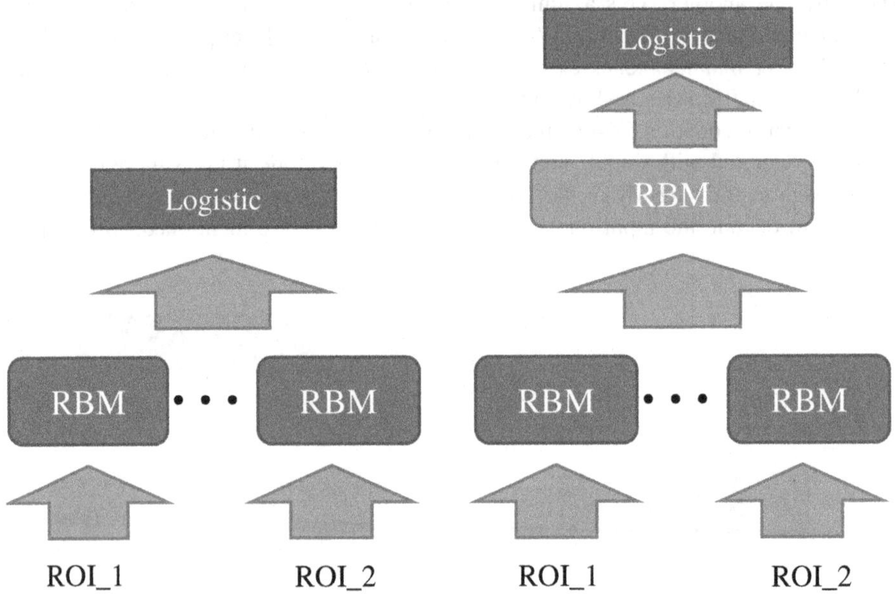

Fig. 4. Decoding process in which the first RBMs are divided into each ROI region

3.1 Experiment Conditions

Decoding experiments for simple motion have been executed for check of validity of DBN-based decoding process, where stimuli were presented on stimuli task blocks. The target data is obtained from two healthy subjects who gave written informed consent.

The experiments are done based on a conventional block design. The design is used in sample experiment of Brain Decoding Tool Box (BDTB) [6]. The hand motion in this task is motions of rock-paper-scissors. The subjects change the hand motion on every 20s after first 20s rest according to instructions on screen. There exist 6 hand motion changes and rest state of 20s in the final block in each run. Each subject conducts 10 runs. During the experiments, EPI data of whole brain is obtained (FOV: 192mm, 64x64 matrix, 3x3x3 mm, 50slices, TE: 50ms, FA 90 degrees, TR: 5s).

Before extraction of feature for classification process from EPI data, preprocessing is executed for all slices. The target ROI of this experiment means M1 and cerebellum (CB), which the regions is defined with SPM5 and WFU PickAtlas [7 - 9] in advance. These can be executed on online processing. The target voxels from the 2 ROI are selected based on t-values calculated by all task data. The number of input voxels in (M1, CB) for 2 subjects is (54, 46) and (83, 117), respectively. The input values mean standardized difference values between target voxel value and rest state voxel value ([0, 1]). The first images of each block are dropped from input data.

The evaluation of decoding result is done by cross validation process in which there exist 18 training blocks and 2 test blocks of each hand motion. The decoding experiments are executed by 2 approaches. First is to classify the hand motion based on hand-crafted feature (batch processing). In this case, the input values are translated by Z score calculated by all run data. The preprocessing finally calculates the average values of standardized values in each block. The classification is executed by SVM which is processed by LibSVM.

In the other approach, the features of input values are constructed by the DBN. In this case, the decoding process is done by 4 types of the DBN. The first is constructed by one RBM and logistic regression. The second is constructed by 3 layers DBN which has two RBMs and logistic regression. As shown in Fig.3, the input layers of the first and the second DBMs deal with 2 ROI values. The third and fourth DBM divide the input values into two ROIs (M1 and CB) as shown in Fig.4. The third is two layers which have 2 RBMs and logistic regression. The fourth is 3 layers which has 3 RBMs and logistic regression. The evaluation of experiments with the DBNs is processed by accuracy of decoding results and computational time in which the DBN is constructed by C#.

3.2 Decoding Results

The decoding experiments as mentioned in 3.1 are done to check the validity of the DBN compared with the conventional decoding process. The accuracy of decoding results based on the conventional process with batch preprocessing is 81.7%, 75%, and 81.7% for only M1 feature, CB feature, and M1 and CB features, respectively.

The accuracy of decoding results with the DBN methods is shown in Table 1. As mentioned in 3.1, the first and the second conditions mean 'whole' input style with 2 and 3 layers, respectively. The third and fourth in 3.1 mean 'divided' input style with 2 and 3 layers, respectively. The accuracy of these methods shows 79.1%, 60.9%, 82.5%, and 67.0%, respectively. The accuracy of the DBN methods with only M1 and CB values shows 82.9% and 75.1%, respectively. These results show no significant differences between the conventional processes and the DBN methods.

The computational time of the DBN methods in which the first RBM layer uses whole voxel values (100 voxels) shows 118s. The time of the divided RBM layers methods shows 41s (46 voxels) and 47s (54 voxels), respectively.

Table 1. Accuracy of decoding results based on each DBN methods

Input style	Layer	Accuracy (%)
Whole	2	79.1
Whole	3	60.9
Divided	2	82.5
Divided	3	67.0

3.3 Discussion

As shown in 3.2, the accuracy of all decoding results shows appropriate results. The decoding results of the conventional process show the almost same accuracy of the previous works in [6]. The accuracy of the DBN methods with 2 layers is comparable results of the conventional results. From this result, online training based on the DBN can extract the features of fMRI data as same as the batch training process. That is, the DBN methods show the possibility of the online decoding process for fMRI task data. Although there is no significant difference between different layers in the DBN methods because of lack of subjects, the accuracy of the methods with 3 layers is lower than that of 2 layers. The RBM can be considered as abstraction process from visible unit values to hidden unit values. Therefore, the second layer of the DBNs abstracts the enough features which can classify the subject's task. That is, the second layers in DBN probably delete the features for classification. The previous work [10] discussed the same situations that RBM is appropriate for the complex classification problem.

The purpose of this comparison results means difference between batch training and online training. Therefore, these experiments satisfy the input conditions. That is, the input values for the DBNs use the difference values between rest state and task state. The difference values in these experiments are useful features for classification because the values of each task exist near constant value. That is, the classification with only the difference values is partially possible although the accuracy is lower than those in these experiments. The reason is that this experiment means simple classification task as above mentioned. Therefore, even the feature extraction based the DBN works well for this experiment situation, the DBN is suitable for more complex situation which needs Multi Voxel Pattern Analysis as discussed in the previous work [10].

The computational time of training process based on divided input style with 2 layers is decrease by about 30 % compares with that of whole input style. Although there is no significance difference of the accuracy, the accuracy of the divided style is higher than that of the whole style. These results suggest that the assumption mentioned in 3.3 is valid for this experiment conditions. That is, it is appropriate for feature extraction of fMRI to divide the target voxel values to each ROI. Although this experiment uses only 2 ROI regions, it is possible to decrease computational cost of the DBN methods based on the dividing input style. The extension of this assumption means convolutional DBN [11] which consider only neighbor voxels.

From these results, the decoding process based on the DBN is appropriate for online training, which these processes are useful items for interactive decoding tasks.

4 Conclusion

The decoding process based on the DBN is constructed for online training in feature extraction. The low level feature of the DBN is calculated for each ROI region based on online RBM training. In order to check the validity of decoding process based on DBN, the decoding experiments of hand motion are executed compared with the conventional decoding process. The accuracy of decoding result based on DBN is comparable to that of the conventional process. Although decoding process does not need complex feature of input voxels because of the easy experiment task, the divided input style for DBN contributes to decreasing the computational cost of training without loss of accuracy. The decoding process based on the DBN achieves online training process for feature extraction and get useful units for interactive decoding experiments for each subject.

References

1. Kamitani, Y., Tong, F.: Decoding the visual and subjective contents of the human brain. Nat. Neurosci. 8(5), 679–685 (2005)
2. Yamashita, O., Sato, M.A., Yoshioka, T., Tong, F., Kamitani, Y.: Sparse estimation automatically selects voxels relevant for the decoding of fMRI activity patterns. Neuroimage 42(4), 1414–1429 (2008)
3. Haxby, J.V., Gobbini, M.I., Furey, M.L., Ishai, A., Schouten, J.L., Pietrini, P.: Distributed and overlapping representations of faces and objects in ventral temporal cortex. Science 293, 2425–2430 (2001)
4. Bengio, Y., Lamblin, P., Popovici, D., Larochelle, H.: greedy layer-wise training of deep networks. Advances in Neural Information Processing Systems 19, 153–160 (2007)
5. Hinton, G.E., Osindero, S., Teh, Y.: A fast learning algorithm for deep belief nets. Neural Computation 18(7), 1527–1554 (2006)
6. Brain Decoder Toolbox (BDTB), `http://www.cns.atr.jp/dni/download/brain-decoder-toolbox/`
7. Lancaster, J.L., Summerln, J.L., Rainey, L., Freitas, C.S., Fox, P.T.: The Talairach Daemon, a database server for Talairach Atlas Labels. NeuroImage 5, S633 (1997)
8. Lancaster, J.L., Woldorff, M.G., Parsons, L.M., Liotti, M., Freitas, C.S., Rainey, L., Kochunov, P.V., Nickerson, D., Mikiten, S.A., Fox, P.T.: Automated Talairach atlas labels for functional brain mapping. Hum. Brain Mapp. 10, 120–131 (2000)

9. Maldjian, J.A., Laurienti, P.J., Kraft, R.A., Burdette, J.H.: An automated method for neuroanatomic and cytoarchitectonic atlas-based interrogation of fmri data sets. NeuroImage 19, 1233–1239 (2003) (WFU Pickatlas, version 3.04)
10. Schmah, T., Hinton, G., Zemel, R.S., Small, S.L., Strother, S.C.: Generative versus discriminative training of RBMs for classification of fMRI images. Advances in Neural Information Processing Systems 21, 1409–1416 (2009)
11. Lee, H., Grosse, R., Ranganath, R., Ng, A.Y.: Convolutional deep belief networks for scalable unsupervised learning of hierarchical representations. In: Proceedings of the Twenty-Sixth International Conference on Machine Learning, ICML 2009 (2009)

Exploiting No-SQL DB for Implementing Lifelog Mashup Platform

Kohei Takahashi, Shinsuke Matsumoto,
Sachio Saiki, and Masahide Nakamura

Kobe University,
1-1 Rokkodai-cho, Nada-ku, Kobe, Hyogo 657-8501, Japan
{koupe@ws.cs,shinsuke@cs,sachio@carp,masa-n@cs}.kobe-u.ac.jp

Abstract. To support efficient integration of heterogeneous lifelog service, we have previously proposed and implemented a *lifelog mashup platform* consisting of the *lifelog common data model (LLCDM)* and the *lifelog mashup API (LLAPI)* to access the standardized data. The LLCDM has standardized columns which is application-independent. And it has application-specific data (i.e. JSON format text of API response of a lifelog service) in the `<content>` column as a plain text. But because the LLCDM repository is implemented using the relational database, we can't access to the `<content>` column data directory, and select out a particular field of it via the LLAPI. To cope with these problems, we exploited the *lifelog mashup platform* with the document-oriented No-SQL database MongoDB for the LLCDM repository. And, we conduct a case study developing an application of retrieving Twitter's posts involving URLs.

Keywords: lifelog, mashup, no-SQL, mongoDB, web services, api.

1 Lifelog Mashup

1.1 Lifelog Services and Mashups

Lifelog is a social act to record and share human life events in open and public form [1]. Various lifelog services currently appear in the Internet. By them, various types of lifelogs are stored, published and shared. For example, *Twitter*[2] for delivering tweets, *Flickr*[3] for storing your photos, *foursquare*[4] for sharing the "check-in" places, blog for writing diary.

Mashup is a new application development approach that allows users to aggregate multiple services to create a service for a new purpose [5]. For example, integrating Twitter and Flickr, we may easily create a photo album with comments (as tweets).

Expecting the lifelog mashup, some lifelog services are already providing APIs, which allow external programs to access the lifelog data. However, there is no standard specification among such APIs or data formats of the lifelog. Figure 1(a) is a data record of Twitter, describing information of a tweet posted by a

K.M. Lee, S.-J. Park, J.-H. Lee (eds.), *Soft Computing in Big Data Processing*,
Advances in Intelligent Systems and Computing 271,
DOI: 10.1007/978-3-319-05527-5_5, © Springer International Publishing Switzerland 2014

user "koupetiko" on 2013-07-05. Figure 1(b) shows a data record retrieved from SensorLoggingService, representing a various sensor's values of user "koupe" on 2013-07-05. We can see that data schema of the two records are completely different. Although both are in the JSON (JavaScript Object Notation), there is no compatibility. Note also that these records were retrieved by different ways by using proprietary APIs and authentic methods. Thus, the mashup applications are developed in an ad-hoc manner, as shown in Figure 2.


```
{
 "created_at": "Fri Jul 05 03:45:35 +0000 2013",
 "id": 353155450876854300,
 "text": "It's sunny, but Im writing a paper in my room...",
 "source": "<a href="http://krile2.star....Krile2</a>",
 ....
 "user": {
  "id": 48441032,
  "screen_name": "koupetiko",
 ....
 "geo": {
  "type": "Point",
  "coordinates": [ 34.72579115, 135.23622368 ]
 },
 "coordinates": { ....
}
```


(a) A data record of Twitter


```
{
 Time: "12:46:57",
 ....
 User: "koupe",
 Weather: "Sunny",
 TempF: 76.73,
 Humidity: 18,
 Brightness: 310,
 Temperature: 26.5,
 ....
 Id: 238361,
 Date: "2013-07-05",
 Light: 81,
 Logger: "PerlSensorLoggerClient.pl",
 Host: "cs27hnspc", ....
}
```


(b) A data record of Sensor Logging Service

Fig. 1. Data of two different lifelog services

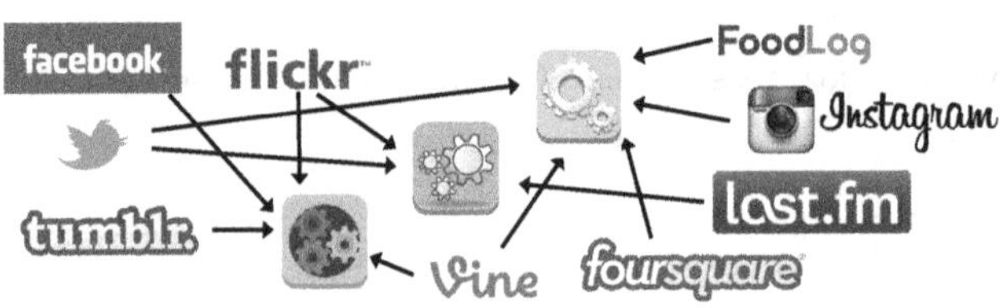

Fig. 2. Conventional Approach of Lifelog Mashup

2 Previous Work

2.1 Lifelog Mashups Platform

To support efficient lifelog mashup, we have proposed a *lifelog mashup platform* [6] previously . The platform consists of the *lifelog common data model (LLCDM)* and the *lifelog mashup API (LLAPI)*, as shown in Figure 3.

Table 1. Common data schema of LLCDM

perspective	data items	Description	Instance
WHEN	`<date>`	Date when the log is created (in UTC)	2013-07-05
	`<time>`	Time when the log is created (in UTC)	12:46:57
	`<epoch>`	UNIX Time (sec) when the log is created (in UTC)	1373028417
WHO	`<user>`	Subjective user of the log	koupe
	`<party>`	Party involved in the log	shinsuke_mat
	`<object>`	Objective user of the log	masa-n
WHERE	`<latitude>`	Latitude where the log is created	34.72631
	`<longitude>`	Longitude where the log is created	135.23532
	`<altitude>`	Altitude where the log is created	141
	`<address>`	Street address where the log is created	1-1, Nada, Kobe
	`<name>`	place name where the log is created	Kobe University
HOW	`<application>`	Service/application by which the log is created	Flickr
	`<device>`	Device with which the log is created	Nikon D7000
WHAT	`<content>`	Contents of the log(whole original data)	`<photo id=".." owner=".." title=".."`
	`<ref_schema>`	URL references to external schema	http://www.flickr.com/services/api/
WHY	n/a	n/a	n/a

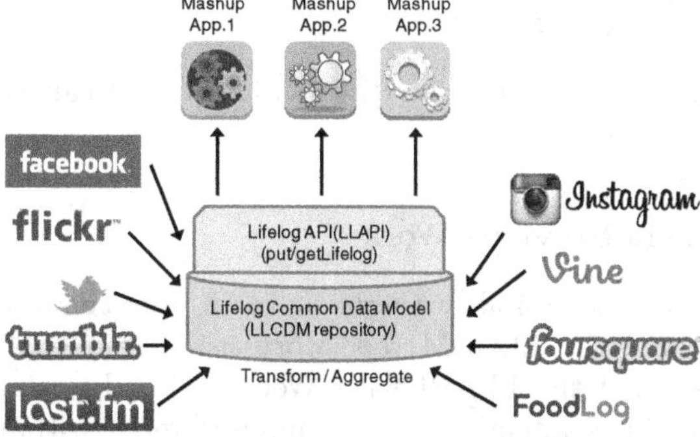

Fig. 3. Proposed Lifelog Mashup Platform [6]

2.2 Lifelog Common Data Model

The data stored in heterogeneous lifelog services are transformed and aggregated in the LLCDM, which is an application-neutral form among the lifelog services.

Table 1 shows the data schema of the LLCDM. We arranged the data items which lifelog records should prepare from the viewpoints of what, why, who, when, where and how. Then, we defined the "common data store" which does not depend on specific service and application.

2.3 Lifelog API

The LifeLog Mashup API (*LLAPI*) is for searching and retrieving lifelog data conforming to the LLCDM. The following shows an API that returns lifelog data matching a given query. Using `getLifeLog()`, heterogeneous lifelogs can be accessed uniformly without proprietary knowledge of lifelog services.

```
getLifeLog(s_date, e_date, s_time, e_time, user, party, object, location,
                                                   application, device, select)
Parameters:
 s_date       : Query of <date>     object       : Query of <object>
 e_date       : Query of <date>     location     : Query of <location>
 s_time       : Query of <time>     application  : Query of <application>
 e_time       : Query of <time>     device       : Query of <device>
 user         : Query of <user>     select       : List of items to be selected
 party        : Query of <party>
```

This is implemented as a query program wrapping an SQL statement. The following SQL statement is published by the `getLifelog()` method.

```
SELECT * FROM Lifelog WHERE
  Date BETWEEN $s_dateq AND $e_dateq
  AND Time BETWEEN $s_timeq AND $e_timeq
  AND UserID LIKE $userq AND Application LIKE $applicationq
  AND Device LIKE $deviceq
```

LLAPI is published as Web service (REST, SOAP) and can be invoked from various platforms.

2.4 Limitations in Previous Work

The limitation is the accessibility to the data in `<content>` column. In [6], we have implemented the LLCDM repository using relational database (i.e. MySQL), and deployed the LLAPI with Web service. In this way, the data of `<content>` column is intentionally left uninterpreted within the LLCDM. Actually, it is stored as plain text in spite of structured data (e.g. xml, json). So, we can't access specified fields in the `<content>` column of lifelog data by queries., although, it is possible to some extent by using text matching. Nevertheless, we can't "select" the specified fields of `<content>` column just like other fields of the LLCDM. For example, it is impossible to retrieve lifelogs of `content.entities.urls` having any items, or to retrieve only `date`, `time`, `content.user.screen_name` columns of each lifelogs, etc. Thus, application developers have to parse `<content>` data in their applications.

The fewer required items in `<content>`, this brings more wasteful development cost, traffic, and the worse performance.

2.5 Research Goal and Approach

Our interest here is to achieve flexible query to the data of `<content>` column.

To achieve this goal. We put all the lifelog data in a document-oriented NoSQL database, instead of having a relational Database. Using this, we aim to flexible access to the data of `<content>` column. But the data except `<content>` column should be fixed according to the LLCDM schema. So, we re-engineer the LLAPI as facade of the database, and it to be used as SQL-like.

3 Exploiting No-SQL DB for Implementing Lifelog Mashup Platform

3.1 Using MongoDB for Lifelog Mashup

To achieve the goals in Section 2.5, we introduce the MongoDB to manage the LLCDM repository (See Figure 3.). MongoDB is a schema less document oriented database developed by 10gen and an open source community [7]. MongoDB was designed to provide both the speed and scalability of key-value datastores as well as the ability to customize queries for the specific structure of the data [8]. In MongoDB, the terms "collection" and "document" are used instead of "table" and "row" in SQL. It has the following features.
Pros

P1: Document-Oriented Storage
MongoDB stores documents as BSON (Binary JSON) objects, which are binary encoded JSON like objects. BSON supports nested object structures with embedded objects and arrays like JSON does [7].

P2: Full Index Support
MongoDB indices are explicitly defined using an ensureIndex call, and any existing indices are automatically used for query processing [9].

P3: High Availability, Easy Scalability
MongoDB supports automatic sharding, distributing documents over servers [9]. And it supports replication with automatic failover and recovery, too. Also MongoDB supports MapReduce.

Cons

C1: No Transaction
MongoDB has no version concurrency control and no transaction management [10]. However, atomic operations are possible within the scope of a single document.

C2: No JOIN
MongoDB doesn't support joins. So, some data is denormalized, or stored with related data documents to remove the need for joins [11].

To use MongoDB for the Lifelog Mashup Platform, P1 and P2 match highly the our purpose, because in the LLCDM repository, data in `<content>` column has various structure and various kinds of elements depends on their application. But, it can't force the data in defined formats or types of each columns, and the columns except `<content>` must be in the defined format in the LLCDM. So, it is necessary to take measures for this policies. By doing this, we will achieve SQL-like searching on all columns of the LLCDM. Also, P3 is very favorable for our future work which is capacity for bigdata.

On the other hand, about C1, most of operations are read and write on the Lifelog Mashup Platform. So, we consider there is no special trouble with C1.

Next, about C2, in data modeling of MongoDB, when entities have "contains" relationships, it is recommended to embedding the child in the parent.

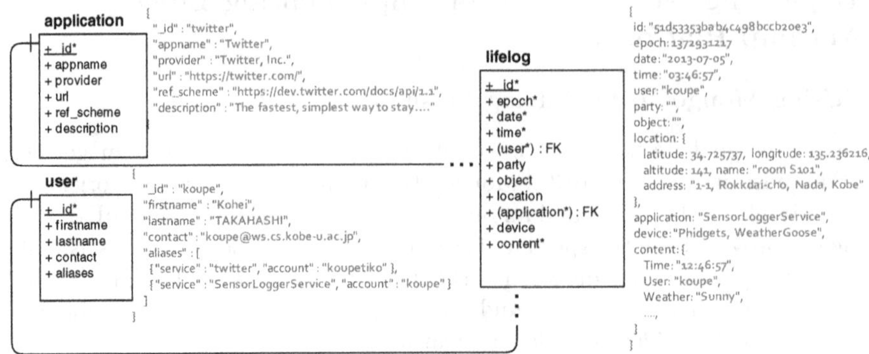

Fig. 4. ER diagram for the LLCDM repository

When entities have one-to-many relationships, it is recommended to embedding or referencing. The embedding is recommended in the case of one-to-many relationships where the "many" objects always appear with or are viewed in the context of their parent. In the new implementation, we comply with this.

3.2 Implementing LLCDM with MongoDB

Considering the above, we have re-designed the data model of the LLCDM for using MongoDB. Figure 4 shows the proposed ER diagram of the LLCDM. A box represents an entity (i.e., table, but it's called collection in MongoDB.), consisting of an entity name, a primary key, foreign key, and attributes. We enumerate instances beside each entity in JSON format to support understanding. A line represents a relationship between entities, where +—· · · denotes a reference relationship. The underlined attributes of entities are primary keys. The attributes in brackets of entities in a relationship are foreign keys. An attribute with a asterisk is NOT Nullable. The diagram consists of the following three collections.

(a) application: In this collection, we manage the applications from which we retrieve the lifelog data record. The application information is useful to identify the types of lifelog data. Therefore, we consider it efficient to manage it in a separate collection. The attributes of this collection are ID, Application name, Provider, URL, Reference URL and Description.

(b) user: This collection manages the user information, consisting of ID, User name, Contact information, Aliases to application accounts. The reason why we provide this collection is that the user information is commonly attached in various lifelog data, and is one of the most frequently used information in the mashup. Also, Aliases achieves to associate various applications' accounts or multiple accounts in the same service with an actual person. This was one of the remaining subjects of previous work [6].

(c) **lifelog:** This is the main collection of the LLCDM. All retrieved lifelog data consisting every item of the LLCDM is stored. As mentioned in Section 2.2 and Table 1, lifelog has each data item arranged from the viewpoints of what, why, who, when, where and how. In this regard, `<ref_schema>` is nothing in lifelog because it is in application.

3.3 Implementing LLAPI with MongoDB

We have implemented `putLifelog()` method and `getLifelog()` method. Since MongoDB is a No-SQL database, the LLAPI must take a role of a part of a relational database (i.e., data type check, data format check, key constraint, and so on.). So, the `putLifelog()` method as a facade of the LLCDM repository. Calling the `putLifelog()` method with parameters corresponding attributes of the entity, the parameters are validated based on the LLCDM and insert to MongoDB. Thus, objects inserted via `putLifelog()` method are guaranteed their formats in the LLCDM schema definition.

Furthermore, SQL-like database retrieval must be kept as same as in previous work. And to achieve more flexible retrieving, we have implemented the `getLifelog()` method as follows.

```
getLifelog([s_date, e_date, s_time, e_time, s_term, e_term, user,
  party, object, s_alt, e_alt, s_lat, e_lat, s_long, e_long,
  loc_name, address, application, device, content, select, limit,
  order, offset])

Parameters:
  s_date, e_date      : Query of <date>
  s_time, e_time      : Query of <time>
  s_term, e_term      : Query of <epoch>
  user, party, object: Query of <user>,<party>,<object>
  s_alt, e_alt        : Query of <location.altitude>
  s_lat, e_lat        : Query of <location.latitude>
  s_long, e_long      : Query of <location.longitude>
  loc_name            : Query of <location.name>
  address             : Query of <location.address>
  application         : Query of <application>
  device              : Query of <device>
  content             : Query/ies of <content>
  select              : List of items to be selected.
  limit               : Limit of retrieved items.
  order               : Order of retrieved items.
  offset              : Skip retrieved items.
  tz                  : Query to adjust date,
                         time, term parameters
```

Table 2. Comparison of Functions

LLAPI	date	time	term	location	application	device	content	select	limit	order	offset	timezone
old(mySQL)	√	√		√	√	√		√				
new(Mongo)	√	√	√	√	√	√	√	√	√	√	√	√

And this method publishes the following MongoDB Query Language. In the case of null parameters, the corresponding query would be nothing (i.e. All parameters were null, `getLifelog()` returns all lifelogs in the LLCDM.). Also, to publish the query, `date`, `time`, `term` parameters are adjusted to UTC time based on `tz` parameter. The `$contentq` is an array of queries to `<content>` column (e.g. "content.temperature $gte 25, content.humidity $lt 40"). The `$selectq` is an array of items to be selected (e.g. "date, time, content.temperature").

```
>db.lifelog.find(
  {
   date: {$ge: $s_dateq, $le: $e_dateq},
   time: {$ge: $s_timeq, $le: $e_timeq},
   epoch: {$ge: $s_termq, $le: $e_termq},
   user: $userq,
   party: $partyq,
   object: $objectq,
   location.altitude: {$ge: $s_altq, $le: $e_altq},
   location.latitude: {$ge: $s_latq, $le: $e_latq},
   location.longitude: {$ge: $s_longq, $le: $e_longq},
   location.name: /$loc_nameq/,
   location.address: /$addressq/,
   application: $applicationq,
   device: $deviceq,
   $contentq[0],
   $contentq[1],
   ...
  }, {
   $selectq[0]: 1,
   $selectq[1]: 1,
   ...
  }
 ).limit($limitq).skip($offsetq).sort({$orderq})
```

Also, Table 2 shows the comparison of functions between the old prototype and the new implementation. In this table, a check mark means an available parameter of each LLAPI. As shown Table 2, the new implementation of the LLAPI is more flexible than the old prototype.

We implemented the LLAPI in the Java language. We used Morphia[12] OR-mapper for marshaling tuples into objects. And for validating objects, we used Java Validation API (Hibernate Validator (JSR303) Reference Implementation for Bean Validation.[13]).

Furthermore, we used JAX-RS (Jersey: JAX-RS (JSR 311) Reference Implementation for building RESTful Web services [14].) to provide RESTful API.

Now that the LLAPI can be accessed by the REST Web service protocol. Figure 5 shows a response of the LLAPI. We can see that a Twitter data record describing a tweet posted by a user "koupetiko" on 2013-05-01 has been retrieved.


```
{
  count: 19,
  result: [
    {
      id: "51d531d1bab4c498bccb1f63",
      date: "2013-05-01",
      time: "13:10:00",
      user: "koupetiko",
      party: "",
      object: "",
      application: "twitter",
      device: "<a href="krile...">Krile2</a>",
      content: {
        source: "<a href="http://krile...">Krile2</a>",
        created_at: "Wed May 01 13:10:00 +0000 2013",
        text: "What's a bad machine...",
          :
        user: {
          profile_image_url_https: "https://..._normal.gif",
          screen_name: "koupetiko",
          ...
        },
        ...
      }
    },
    {
      id: "51d531d1bab4c498bccb1f64",
      date: "2013-05-01",
      time: "13:08:05",
      ...
    },
    ...
  ]
}
```


Fig. 5. a response of getLifelog

4 Case Study

Using the proposed method, we develop a practical Web application *My Tweeted URLs*. The application *My Tweeted URLs* provides a simple viewer of one's tweets involving urls and their web pages' thumbnails. And the thumbnails link their web pages. For retrieving and processing the lifelogs of tweets, we have used jQuery. Also, for obtaining thumbnails, we have used public API of WordPress.com. The total lines of code is just 96, including the code of html and jQuery. Figure 6 is the screenshot of this application.

We have implemented this application so easily because we only had to obtain the thumbnails and show the tweets of retrieved lifelogs from the LLCDM. This process is the essential process of this application.

If we use old prototype, to implement this, we would have to do some wasted jobs. First, we retrieve all lifelog data in the target period from the LLCDM, and parse their `<content>` column data in JSON format text to JavaScript object. After that, check each data if its tweet has "http". At last, we change over to essential process of obtaining the thumbnails and showing it with tweets.

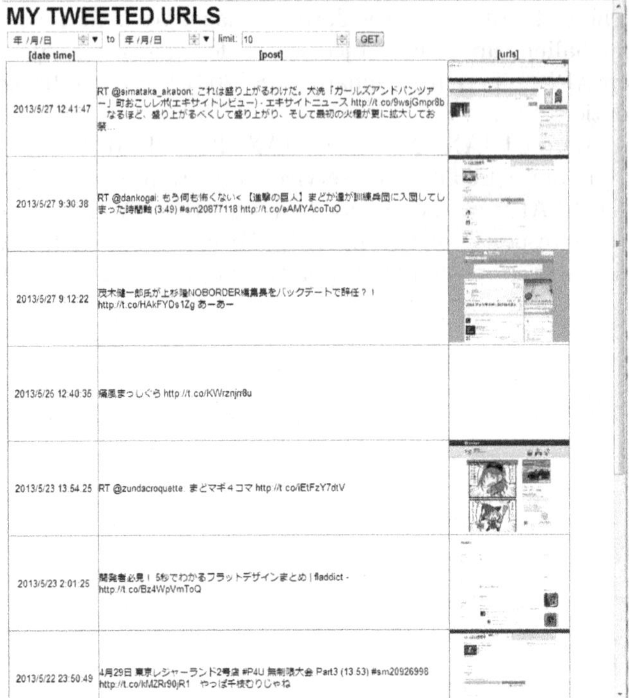

Fig. 6. Screenshot of "My Tweeted URLs"

5 Conclusion

In this paper, to improve the accessibility of the data, we have re-engineered the lifelog mashup platform [6], consisting of the LLCDM (LifeLog Common Data Model) and the LLAPI (LifeLog Mashup API). Specifically, we exploited the No-SQL database for the LLCDM repository to achieve flexible query to `<content>` column data. As a case study, we developed the lifelog application "My Tweeted URLs", and we made sure that the proposed method would reduce the development effort and complexity of source code of the lifelog application.

Our future work includes performance improvement, and evaluate the capacity for bigdata. We also plan to study potentials of lifelog mashups for business and education.

Acknowledgments. This research was partially supported by the Japan Ministry of Education, Science, Sports, and Culture [Grant-in-Aid for Scientific Research (C) (No.24500079), Scientific Research (B) (No.23300009)].

References

1. Trend Watching.com: Life caching – an emerging consumer trend and related new business ideas, `http://trendwatching.com/trends/LIFE_CACHING.htm`
2. Twitter, `http://twitter.com/`

3. Flickr, `http://www.flickr.com/`
4. foursquare, `http://foursquare.com/`
5. Lorenzo, G.D., Hacid, H., Young Paik, H., Benatallah, B.: Data integration in mashups. ACM 38, 59–66 (2009)
6. Shimojo, A., Matsuimoto, S., Nakamura, M.: Implementing and evaluating life-log mashup platform using rdb and web services. In: The 13th International Conference on Information Integration and Web-based Applications & Services (iiWAS 2011), pp. 503–506 (December 2011)
7. Padhy, R.P., Patra, M.R., Satapathy, S.C.: Rdbms to nosql: Reviewing some next-generation non-relational databases. International Journal of Advanced Engineering Science and Technologies 11(1), 15–30 (2011)
8. Bunch, C., Chohan, N., Krintz, C., Chohan, J., Kupferman, J., Lakhina, P., Li, Y., Nomura, Y.: Key-value datastores comparison in appscale (2010)
9. Cattell, R.: Scalable sql and nosql data stores. SIGMOD Rec. 39(4), 12–27 (2011)
10. Milanović, A., Mijajlović, M.: A survey of post-relational data management and nosql movement
11. 10gen, Inc.: Mongodb, `http://www.mongodb.org/`
12. Morphia, `https://github.com/mongodb/morphia`
13. Hibernate-Validator, `http://www.hibernate.org/subprojects/validator.html`
14. Jersey, `http://jersey.java.net/`

Using Materialized View as a Service of Scallop4SC for Smart City Application Services

Shintaro Yamamoto, Shinsuke Matsumoto,
Sachio Saiki, and Masahide Nakamura

Kobe University,
1-1 Rokkodai-cho, Nada-ku, Kobe, Hyogo 657-8501, Japan
{shintaro@ws.cs, shinsuke@cs,
sachio@carp, masa-n@cs}.kobe-u.ac.jp

Abstract. Smart city provides various value-added services by collecting large-scale data from houses and infrastructures within a city. However, it takes a long time and man-hour and needs knowledge about big data processing for individual applications to use and process the large-scale raw data directly. To reduce the response time, we use the concept of *materialized view* of database, and materialized view to be as a service. And we propose materialized view to be as as service (*MVaaS*). In our proposition, a developer of an application can efficiently and dynamically use large-scale data from smart city by describing simple data specification without considering distributed processes and materialized views. In this paper, we design an architecture of MVaaS using MapReduce on Hadoop and HBase KVS. And we demonstrate the effectiveness of MVaaS through three case studies. If these services uses raw data, it needs enormous time of calculation and is not realistic.

Keywords: large-scale, house log, materialized view, high-speed and efficient data access, MapReduce, KVS, HBase.

1 Smart City and Scallop4SC

1.1 Smart City and Services

The principle of the smart city is to gather data of the city first, and then to provide appropriate services based on the data. Thus, a variety of data are collected from sensors, devices, cars and people across the city. A smart city provides various value-added services, named *smart city services*, according to the situation by big data within a city. Promising service fields include energy saving [1], traffic optimization [2], local economic trend analysis [3], entertainment [4], community-based health care [5], disaster control [6] and agricultural support [7].

The size and variety of gathered data become huge in general. Velocity (i.e., freshness) of the data is also important to reflect real-time or latest situations and contexts. Thus, the data for the smart city services is truly *big data*.

Due to the limitation of storage, the conventional applications were storing only necessary data with optimized granularity. Therefore, the gathered data was application-specific, and could not be shared with other applications.

K.M. Lee, S.-J. Park, J.-H. Lee (eds.), *Soft Computing in Big Data Processing*,
Advances in Intelligent Systems and Computing 271,
DOI: 10.1007/978-3-319-05527-5_6,

The limitation of the storage is relaxed significantly by cloud computing technologies. Thus, it is now possible to store various kinds of data as they are, and to reuse the raw data for various purposes. We are interested in constructing a data platform to manage the big data for smart city services.

1.2 Scallop4SC (Scalable Logging Platform for Smart City)

We have been developing a data platform, called *Scallop4SC*, for smart city services [8][9]. Scallop4SC is specifically designed to manage data from *houses*. The data from houses are essential for various smart city services, since a house is a primary construct of a city. In near future, technolgies of smart homes and smart devices will enable to gather various types of house data.

Scallop4SC basically manages two types of house data: *house log* and *house configuration*. The house log is history of values of any dynamic data measured within a smart home. Typical house log includes power consumption, status of an appliance and room temperature. The house configuration is static meta-data explaining a house. Examples include house address, device ID, floor plan and inhabitant names.

Figure 1 shows the architecture of Scallop4SC. For each house in a smart city, a logger measures various data and records the data as house log. The house log is periodically sent to Scallop4SC via a network. Due to the number of houses and the variety of data, the house log generally forms big data. Thus, Scallop4SC stores the house log using *HBase* NoSQL-DB, deployed on top of *Hadoop* distributed processing. On the other hand, the house configuration is static but structural data. Hence, it is stored in *MySQL* RDB to allow complex queries over houses, devices and people.

Scallop4SC API (shown in the middle in Figure 1) provides a basic access method to the stored data. Since Scallop4SC is an application-neutral platform, the API just allows basic queries (see [9]) to retrieve the *raw data*. Application-specific data interpretation and conversion are left for individual applications.

1.3 Introducing Materialized View in Scallop4SC

In general, individual applications use the smart city data in different ways. If an application-specific data is derived from much of raw data, the application suffers from expensive data processing and long processing time. This is because the application-specific data conversion is left to each application. If the application repeatedly requires the same data, the application has to repeat the same calculation to the large-scale data, which is quite inefficient.

To cope with this, we introduced *materialized view* in Scallop4SC, as shown in the lower part of Figure 1 [10]. The application-specific data can be considered as a *view*, which looks up the raw data based on a certain query. The materialized view is constructed as a table, which *caches* results of the query in advance.

Note, however, that the raw data in Scallop4SC is very large, and that we cannot use SQL for HBase to construct the view. Therefore, in [10] we developed a Hadoop/MapReduce program for each application, which efficiently converts the raw data into

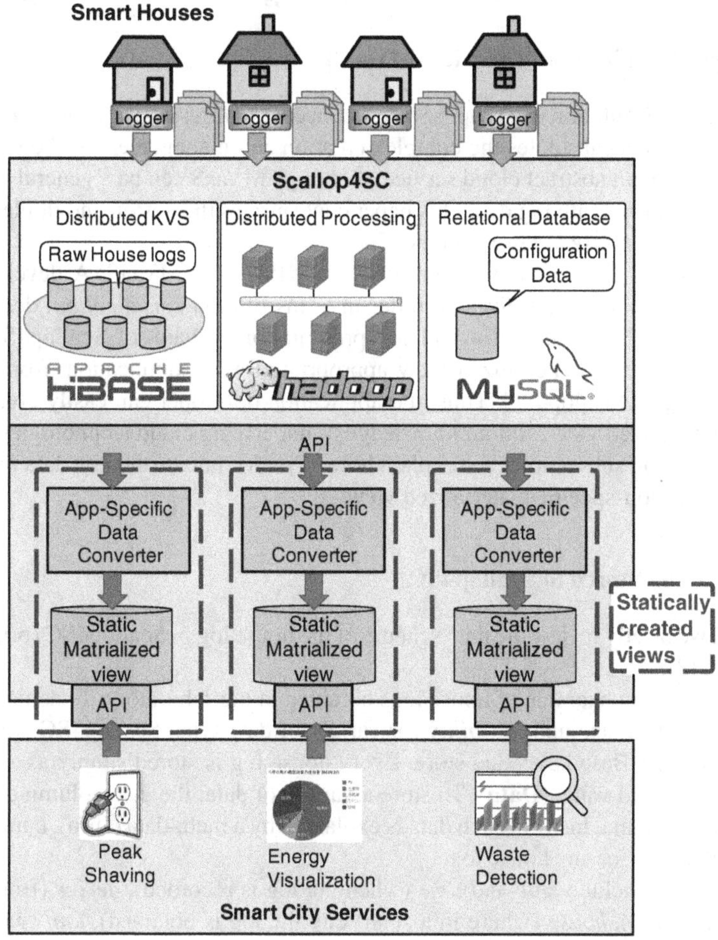

Fig. 1. Scallop4SC with static MVs

application-specific data. The converted data is stored in an HBase table, which is used as a materialized view by the application. Our experiment showed that the use of materialized view significantly reduced computation cost of applications and improved the response time.

A major limitation of the previous research is that the MapReduce program was statically developed and deployed. This means that each application developer has to implement a proprietary data converter by himself. The implementation requires development effort as well as extensive knowledge of HBase and Hadoop/MapReduce. It is also difficult to reuse the existing materialized views for other applications. These are obstacle for rapid creation of new applications.

2 MaterializedView as a Service for Large-Scale House Log

2.1 Materialized View as a Service (MVaaS)

To overcome the limitation, we propose a new concept of *Materialized View as a Service (MVaaS)*. MVaaS encapsulates the complex creation and management of the materialized views within an abstract cloud service. Although MVaaS can be a general concept for any data platform with big data, this paper concentrates the design and implementation of MVaaS for *house log* in Scallop4SC.

Figure 2 shows the new architecture of Scallop4SC with MVaaS. A developer of a smart city application first gives an order in terms of *data specification*, describing what data should be presented in which representation. MVaaS of Scallop4SC then dynamically creates a materialized view appropriate for the application, from large-scale house log of Scallop4SC. Thus, the application developer can easily create and use own materialized view without knowledge of underlying cloud technologies.

In the following subsections, we explain how MVaaS converts the raw data of house log into application-specific materialized view.

2.2 House Log Stored in Scallop4SC

First of all, we briefly review the data schema of the house log in Scallop4SC (originally proposed in [8]).

Table 1 shows an example of house logs obtained in our laboratory. To achieve both *scalability* for data size and *flexibility* for variety of data type, Scallop4SC stores the house log in the HBase key value store. Every house log is stored simply as a pair of key (**Row Key**) and value (**Data**). To store a variety of data, the data column does not have rigorous schema. Instead, each data is explained by a meta-data (**Info**), comprising standard attributes for any house log.

The attributes include *date* and *time* (when the log is recorded), *device* (from what the log is acquired), *house* (where in a smart city the log is obtained), *unit* (what unit should be used), *location* (where in the house the log is obtained) and *type* (for what the log is). Details of device, house and location are defined in an external database of house configuration in MySQL (see Figure 2). A row key is constructed as a concatenation of date, time, type, home and device. An application can *get* a single data (i.e., row) by specifying a row key. An application can also *scan* the table to retrieve multiple rows by *prefix match* over row keys. Note that complex queries with SQL cannot be used to search data, since HBase is a NoSQL database.

Table 1. Raw Data: House Log of Scallop4SC

Row Key	Column Families							
	Data:	Info:						
(dateTtime.type.home.device)		date:	time:	device:	house:	unit:	location:	type:
2013-05-28T12:34:56.Energy.cs27.tv01	600	2013-05-28	12:34:56	tv01	cs27	W	living room	Energy
2013-05-28T12:34:56.Device.cs27.tv01	[power:off]	2013-05-28	12:34:56	tv01	cs27	status	living room	Device
2013-05-28T12:34:56.Environment.cs27.temp3	24.0	2013-05-28	12:34:56	temp3	cs27	Celsius	kitchen	Environment
2013-05-28T12:34:56.Environment.cs27.pcount3	3	2013-05-28	12:34:56	pcount3	cs27	people	living room	Environment
2013-05-28T12:35:00.Device.cs27.tv01	on()	2013-05-28	12:35:00	tv01	cs27	operation	living room	Device
:	:	:	:	:	:	:	:	:

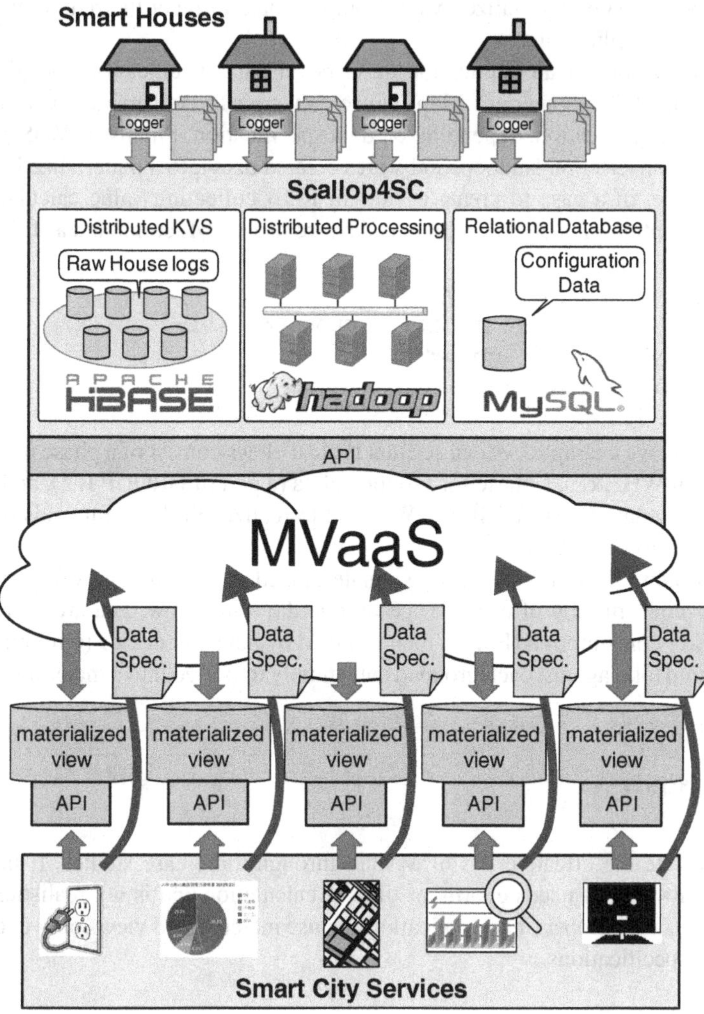

Fig. 2. Extended Scallop4SC

For example, the first row in Table 1 shows that the log is taken at 12:34:56 on May 28, 2013, and that the house ID is cs27 and log type is Energy, and that the deviceID is tv01. The value of power consumption is 600 W. Similarly, the second row shows a status of tv01 where power is off.

We assume that these logs are used as *raw data* by various smart city services and applications.

2.3 Converting Raw Data into Materialized View

In order to create materialized views in response to applications, it needs converting data from raw data and importing the materialized view to create the materialized view.

Here is how to create materialized view from raw data to meet the requirements of a developer of an application.

First, a developer of an application describes a data specification and gives it to MVaaS. Next, MVaaS generates MapReduce converter to convert raw data into data suitable for the application, depending on data specification. Finally, MVaaS practices MapReduce Converter on Hadoop, and stores created data into a materialized view.

For example, in a case to create a view in RDB collecting value calculated total power consumption of each device by one day, we use a SQL command as follows:

```
CREATE VIEW view_name AS
SELECT date,device,SUM(Data) FROM houselog
WHERE type=Energy,unit=W
GROUP BY date,device;
```

From the above example, we can see that typical views consist of a phase for filtering data (a part of WHERE), a phase for grouping data (a part of GROUP BY), and a phase for aggregating data (a part of SUM). We want to realize similar commands for house logs on HBase to be NoSQL.

Figure 3 shows flowchart of creating a materialized view collecting value calculated total power consumption of each device by one day. First, raw data are narrowed to refer to power consumption (Filter), then grouped by data and device (Grouping), next aggregate sum total against each grouped data, finally imported into a materialized view.

3 Case Study

We demonstrate the effectiveness of MVaaS through three case studies. If these services uses raw data, it needs enormous time of calculation and is not realistic. But by using MVaaS, a developer of an application can use materialized views only to describe simple data specifications.

3.1 Power Consumption Visualization Service

The service collects data of power consumption from smart house and visualizes the use of energy from various viewpoints (e.g. houses, towns, devices, current power consumption, passage of past power consumption, etc.). This service is intended to raise user's awareness of energy saving by intuitively showing the current and past usage of energy.

Here, we will examine about one of simple power consumption visualization services as a case study. The service is to visualize power consumption for each device by one hour in one house.

Thus, this application needs logs that type is Energy, unit is W, house is a house (cs27), row key consists time and device, and value is total sum of power consumption, like Table 2.

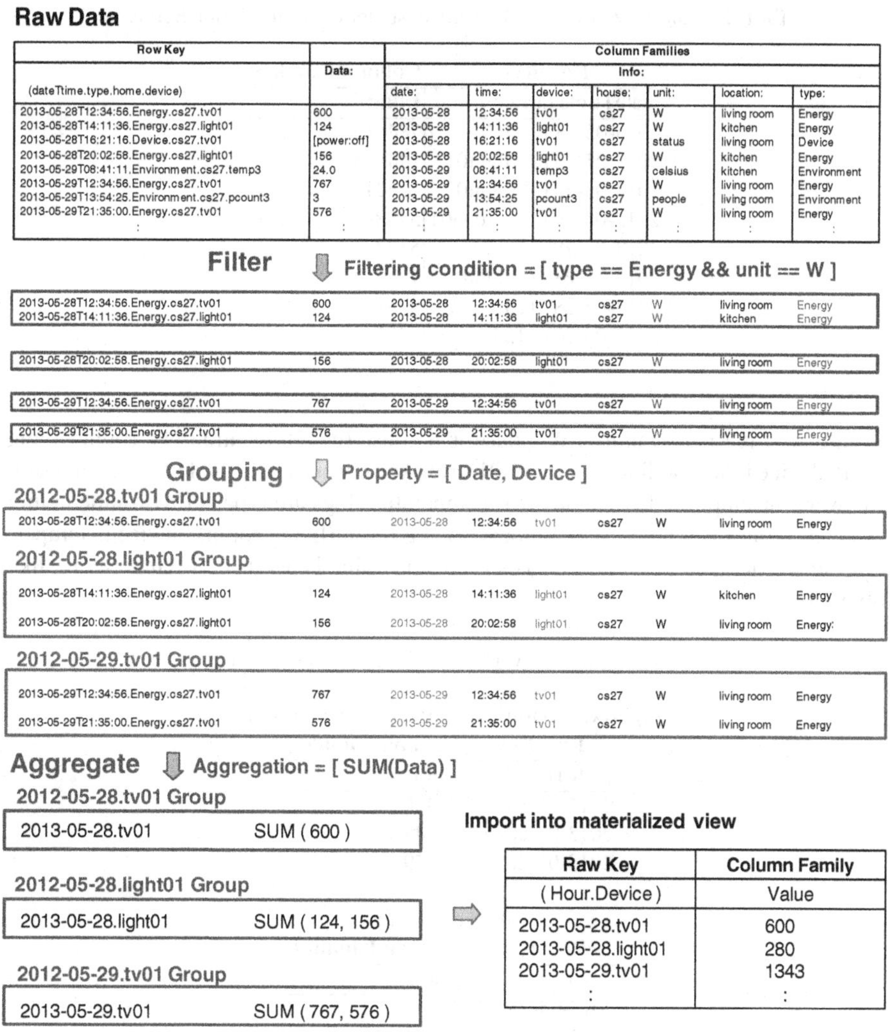

Raw Data

Row Key		Column Families						
	Data:	Info:						
(dateTtime.type.home.device)		date:	time:	device:	house:	unit:	location:	type:
2013-05-28T12:34:56.Energy.cs27.tv01	600	2013-05-28	12:34:56	tv01	cs27	W	living room	Energy
2013-05-28T14:11:36.Energy.cs27.light01	124	2013-05-28	14:11:36	light01	cs27	W	kitchen	Energy
2013-05-28T16:21:16.Device.cs27.tv01	[power:off]	2013-05-28	16:21:16	tv01	cs27	status	living room	Device
2013-05-28T20:02:58.Energy.cs27.light01	156	2013-05-28	20:02:58	light01	cs27	W	kitchen	Energy
2013-05-29T08:41:11.Environment.cs27.temp3	24.0	2013-05-29	08:41:11	temp3	cs27	celsius	kitchen	Environment
2013-05-29T12:34:56.Energy.cs27.tv01	767	2013-05-29	12:34:56	tv01	cs27	W	living room	Energy
2013-05-29T13:54:25.Environment.cs27.pcount3	3	2013-05-29	13:54:25	pcount3	cs27	people	living room	Environment
2013-05-29T21:35:00.Energy.cs27.tv01	576	2013-05-29	21:35:00	tv01	cs27	W	living room	Energy
:	:	:	:	:	:	:	:	:

Filter ⇓ Filtering condition = [type == Energy && unit == W]

2013-05-28T12:34:56.Energy.cs27.tv01	600	2013-05-28	12:34:56	tv01	cs27	W	living room	Energy
2013-05-28T14:11:36.Energy.cs27.light01	124	2013-05-28	14:11:36	light01	cs27	W	kitchen	Energy
2013-05-28T20:02:58.Energy.cs27.light01	156	2013-05-28	20:02:58	light01	cs27	W	living room	Energy
2013-05-29T12:34:56.Energy.cs27.tv01	767	2013-05-29	12:34:56	tv01	cs27	W	living room	Energy
2013-05-29T21:35:00.Energy.cs27.tv01	576	2013-05-29	21:35:00	tv01	cs27	W	living room	Energy

Grouping ⇓ Property = [Date, Device]

2012-05-28.tv01 Group

2013-05-28T12:34:56.Energy.cs27.tv01	600	2013-05-28	12:34:56	tv01	cs27	W	living room	Energy

2012-05-28.light01 Group

2013-05-28T14:11:36.Energy.cs27.light01	124	2013-05-28	14:11:36	light01	cs27	W	kitchen	Energy
2013-05-28T20:02:58.Energy.cs27.light01	156	2013-05-28	20:02:58	light01	cs27	W	living room	Energy

2012-05-29.tv01 Group

2013-05-29T12:34:56.Energy.cs27.tv01	767	2013-05-29	12:34:56	tv01	cs27	W	living room	Energy
2013-05-29T21:35:00.Energy.cs27.tv01	576	2013-05-29	21:35:00	tv01	cs27	W	living room	Energy

Aggregate ⇓ Aggregation = [SUM(Data)]

2012-05-28.tv01 Group

2013-05-28.tv01 SUM (600)

2012-05-28.light01 Group

2013-05-28.light01 SUM (124, 156)

2012-05-29.tv01 Group

2013-05-29.tv01 SUM (767, 576)

⇨ **Import into materialized view**

Raw Key	Column Family
(Hour.Device)	Value
2013-05-28.tv01	600
2013-05-28.light01	280
2013-05-29.tv01	1343
:	:

Fig. 3. Flowchart of converting raw data into materialized view

3.2 Wasteful Energy Detection Service

This service automatically detects wasteful electricity and notifies users using power consumption data and sensors data in a smart house. Furthermore, a user can review the detected waste occurred in the past.

Here, we will examine about one of wasteful energy detection service as a case study. The service is to visualize power consumption of an air conditioner with temperature in the room.

Thus, this application needs two logs. First one is logs about power consumption of an air conditioner. Another is logs about temperature in the room. Power consumption

Table 2. Materialized view of Power Consumption Visualization Service

Row Key	**Column Families**
(dateTtime.device)	**Data:**
2013-05-28T10.tv01	784415
2013-05-28T10.light01	1446
2013-05-28T10.light02	21321
2013-05-29T10.aircon01	54889
2013-05-28T11.tv01	51247
2013-05-28T11.light01	7742
2013-05-28T11.light02	3288
2013-05-28T11.aircon01	65774
:	:

logs are type is Energy, unit is W, house is a house (cs27), location is a room (living room), device is air conditioner (aircon01), row key consists time, and value is total sum of power consumption, like Table 3(a). Temperature logs are type is Environment, unit is Celsius, house is a house (cs27), location is a room (living room), device is temperature sensor (temp1), row key consists time, and value is average of temperature, like Table 3(b).

Table 3. Materialized view of Wasteful Energy Detection Service

(a) Power consumption of the air conditioner

Row Key	**Column Families**
(dateTtime)	**Data:**
2013-05-28T10	75426
2013-05-28T11	11548
2013-05-28T12	21859
:	:

(b) Temperature

Row Key	**Column Families**
(hour)	**Data:**
2013-05-28T10	27.0
2013-05-28T11	23.4
2013-05-28T12	28.2
:	:

3.3 Diagnosis and Improvement of Lifestyle Support Service

This service urges users to improve their lifestyles and advises users on health with analyzing the daily device logs. For example, by using light log in user's house the service advises on time of sleep.

Here, we will examine about one of simple power consumption visualization services as a case study. The service is to visualize sleep period of time from logs. For example, the period of time in which all lights are off with pelple in the room is expected to be a sleep period of time.

Table 4. Materialized view of Diagnosis and Improvement of Lifestyle Support Service

(a) Status of lights

Row Key	**Column Families**
(date.location.device)	**Data:**
2013-05-28.living room.tv01	21:12:44, 21:13:43, 21:14:57,...
2013-05-28.bedroom.light01	23:34:56, 23:35:56, 23:36:57,...
2013-05-29.living room.tv01	20:22:16, 20:23:14, 20:24:17,...
2013-05-29.bedroom.light01	21:46:23, 21:47:24, 21:48:50,...
2013-05-30.living room.tv01	23:56:21, 23:57:21, 23:58:24,...
2013-05-30.bedroom.light01	23:57:23, 23:58:21, 23:59:20,...
:	:

(b) Period of time of people in the room

Row Key	**Column Families**
(date.location)	**Data:**
2013-05-28.living room	18:12:44, 18:13:43, 18:14:57,...
2013-05-28.bedroom	22:34:56, 22:35:56, 22:36:57,...
2013-05-29.living room	19:22:16, 19:23:14, 19:24:17,...
2013-05-29.bedroom	21:36:23, 21:37:24, 21:38:50,...
2013-05-30.living room	16:56:21, 16:57:21, 16:58:24,...
2013-05-30.bedroom	20:57:23, 20:58:21, 20:59:20,...
:	:

Thus, this application needs two logs. First one is logs about status of devices. Another is logs about presence of people in the room. Status of devices logs are type is Device, unit is status, house is a house (cs27), row key consists date, room and device , and value is time in which devices is off, like Table 4(a). Presence of people in the room logs are type is Environment, unit is peolple, house is a house (cs27), row key consists date and room, and value is time in which people exist, like Table 4(b).

In this case study, we thought the case using two materialized views. But this two materialized views can be combined into one. If we combine materialized views, number of data will decrease and it will be easy to visualize, whereas extensibility and reusability will not maintain.

4 Conclusion

In this paper, we proposed an architecture of MVaaS that allows various applications to efficiently and dynamically use large-scale data. The method is specifically applied to a data platform, Scallop4SC, for the large-scale smart city data. In the MVaaS, a developer of an application can efficiently and dynamically access large-scale smart city data only by describing data specifications for application. And we demonstrated the effectiveness of MVaaS through three case studies. If these services uses raw data, it needs enormous time of calculation and is not realistic.

Our future works include implement of MVaaS, as well as an efficient operation method of MapReduce converter.

Acknowledgments. This research was partially supported by the Japan Ministry of Education, Science, Sports, and Culture [Grant-in-Aid for Scientific Research (C) (No.24500079), Scientific Research (B) (No.23300009)].

References

1. Massoud Amin, S., Wollenberg, B.: Toward a smart grid: power delivery for the 21st century. IEEE Power and Energy Magazine 3(5), 34–41 (2005)
2. Brockfeld, E., Barlovic, R., Schadschneider, A., Schreckenberg, M.: Optimizing traffic lights in a cellular automaton model for city traffic. Phys. Rev. E 64, 056132 (2001)
3. IBM: Apply new analytic tools to reveal new opportunities, http://wwiw.bm.com/smarterplanet/nz/en/business_analytics/article/it_business_intelligence.html
4. Celino, I., Contessa, S., Corubolo, M., Dell'Aglio, D., Valle, E.D., Fumeo, S., Krüger, T.: Urbanmatch - linking and improving smart cities data. In: Linked Data on the Web, LDOW 2012 (2012)
5. IBM: eHealth and collaboration - collaborative care and wellness, http://www.ibm.com/smarterplanet/nz/en/healthcare_solutions/nextsteps/solution/X056151Y25842W14.html
6. Hu, C., Chen, N.: Geospatial sensor web for smart disaster emergency processing. In: International Conference on GeoInformatics (Geoinformatics 2011), pp. 1–5 (2011)
7. Cho, Y., Moon, J., Yoe, H.: A context-aware service model based on workflows for u-agriculture. In: Taniar, D., Gervasi, O., Murgante, B., Pardede, E., Apduhan, B.O. (eds.) ICCSA 2010, Part III. LNCS, vol. 6018, pp. 258–268. Springer, Heidelberg (2010)
8. Yamamoto, S., Matumoto, S., Nakamura, M.: Using cloud technologies for large-scale house data in smart city. In: International Conference on Cloud Computing Technology and Science (CloudCom 2012), pp. 141–148 (December 2012)
9. Takahashi, K., Yamamoto, S., Okushi, A., Matsumoto, S., Nakamura, M.: Design and implementation of service api for large-scale house log in smart city cloud. In: International Workshop on Cloud Computing for Internet of Things (IoTCloud 2012), pp. 815–820 (December 2012)
10. Ise, Y., Yamamoto, S., Matsumoto, S., Nakamura, M.: In: International Conference on Software Engineering, Artificial Intelligence, Networking and Parallel/Distributed Computing, SNPD 2013 (2012) (to appear)

Noise Removal Using TF-IDF Criterion for Extracting Patent Keyword

Jongchan Kim[1,3], Dohan Choe[1,3], Gabjo Kim[1,3], Sangsung Park[2,3], and Dongsik Jang[1,3,*]

[1] Division of Industrial Management Engineering, Korea University
[2] Graduate School of Management of Technology, Korea University
[3] 1, 5-ga, Anam-dong Seongbuk-gu, Seoul, 136-701 Korea

Abstract. These days, governments and enterprises are analyzing trends in technology as a part of their investment strategy and R&D planning. Qualitative methods by experts are mainly used in technology trend analyses. However, such methods are inefficient in terms of cost and time for large amounts of data. In this study, we quantitatively analyzed patent data using text mining with TF-IDF used as weights. Keywords and noises were also classified using TF-IDF weighting. In addition, we propose new criteria for removing noises more effectively, and visualize the resulting keywords derived from patent data using social network analysis (SNA).

Keywords: Keyword extraction, Patent analysis, Text mining, TF-IDF.

1 Introduction

It has recently become important for companies and nations to understand the technology trends of the future and forecast emerging technologies while planning their R&D and investment strategies. For such technology forecasting, a qualitative method such as Delphi has been mainly used through the subjective opinions and assessments of specialists. However, these types of qualitative methods have certain drawbacks:

1. Subjective results are deducted in a qualitative manner.
2. Professional manpower in the target field is required.
3. Money and time are wasted.

These disadvantages in qualitative technology trend analyses can be complemented through the use of a quantitative method, which draws an objective result and saves money and time. It also allows non-experts to perform a technology trend analysis. In this paper, we therefore quantitatively analyze the technology trend of carbon fibers based on patent information.

* Corresponding author.

K.M. Lee, S.-J. Park, J.-H. Lee (eds.), *Soft Computing in Big Data Processing*, Advances in Intelligent Systems and Computing 271,
DOI: 10.1007/978-3-319-05527-5_7,

Patent information extracted for an analysis can be divided into two data groups, technical data and bibliographic data. Bibliographic data contains structured data such as the application date, priority date, publication number, registration number, applicant, inventor, international patent classification (IPC) code, the patent classification of each country, the applicant country, designated states, patent citations, and attorneys. Technical data are well-summarized technical ideas including technologies, materials, products, effects, compositions, treatments, processes, functions, usages, technology stages, titles, and abstracts. Unlike bibliographic data, technical data are unstructured. Therefore, experts who fully understand technology descriptions mainly use quantitative analysis methods for technical data. However, a quantitative method is inefficient for analyzing massive amounts of data on technology trends because of the time and effort required. A quantitative method can be used to solve this problem [1]. A text mining method is used to analyze quantitatively unstructured technical data. We can use technical data that has been preprocessed using a text mining method to calculate the weight score, such as term frequency–inverse document frequency (TF-IDF), or analyze technology trends after applying a data mining method such as clustering, pattern exploration, or classification.

We selected keywords in the field of carbon fiber technology using text mining and a TF-IDF measure, and then visualized the technology trend through a weighted social network analysis (SNA) network. We gained objective results quickly and simply using this process of technology trend analysis.

2 Text Mining

Natural Language Processing is a computer technique for performing analytical processing of natural language and understanding its structure and meaning. Although studies on Natural Language Processing have been conducted with the development of the computer, Natural Language Processing still requires further research. Text mining is a technique used to extract meaningful information and value from unstructured text data. By applying a statistical algorithm to text mining, it is possible to obtain more results than in a simple information search. The primary use of text mining is document classification, which is used to sort documents based on a classification rule, document clustering to gather documents according to their correlation and similarity, and information extraction to find significant information in a document [2]. And There are several studies on text analysis [3,4].

In this study, we use text mining to select keywords that represent a technology from each patent document.

3 TF-IDF

TF-IDF is a type of weighting used in information retrieval and text mining. The weighting indicates how important a certain word ‘A’ is in a specific document among all words in the document data. The TF-IDF is the value obtained by multiplying the IDF and TF. The higher the value is, the more important the meaning word 'A' has in a specific document.

tf_A : Frequency of word A in a specific document (normalized)

$$idf_A = \log\frac{N}{df_A} \tag{1}$$

N : Number of all documents

df_A : Frequency of documents that include word A

$$w_A = tf_A \times idf_A \tag{2}$$

$$w_{kj} = \frac{tfidf(A_k, d_j)}{\sqrt{\sum_1^r (tfidf(A_s, d_j))^2}} \tag{3}$$

Term frequency (TF) is a value indicating how frequently word 'A' appears in a specific document. TF has a form normalized by dividing the appearance frequency of word 'A' by the other words. A higher TF value indicates that the word has important information. However, if a certain word frequently emerges in not only a specific document but in many documents, it indicates that the word is not a keyword because it is commonly used. For example, when investigating technology trends, the words 'fiber' and 'carbon' appear in many patent documents collected in the carbon fiber field. However, 'fiber' and 'carbon' cannot explain the technology of each patent. As described in Eq. (1), while document frequency (DF) is the number of documents that word 'A' appears in, inverse document frequency (IDF) is the inverse value of DF. To give the relative weight to TF more than to IDF, a log function is applied to the latter. Since the length of each document differs, the point to consider in TF-IDF is that the TF value can change simply according to the length of the document. Eq. (3) describes a value normalized by considering the difference in length of a respective document. Jeong(2012) and Lee(2009) derived important keywords in the News Corpus and construction field by applying the deformed TF-IDF weight [5, 6].

This study proposes a new threshold to derive a keyword using a normalized TF-IDF weight.

4 Social Network Analysis

A social network analysis is an analysis methodology that consists of multiple nodes and edge linking nodes to analyze and visualize the relationship between people, groups, and data. The structural hole, degree, density, and Euclidean distance are indicators for measuring the cohesion of a network, whereas the centrality, centralization, and reciprocity are characteristics of a network. Social network analyses can be divided into simple networks, networks weighted to the node and edge, and bipartite networks consisting of n numbers of nodes in rows and m numbers of nodes in columns [7]. In this study, we used a weighted network analysis to visually analyze associations between two or more extracted keywords in a certain document.

5 Research Methodology

The research methodology proposed in this study is as follows.

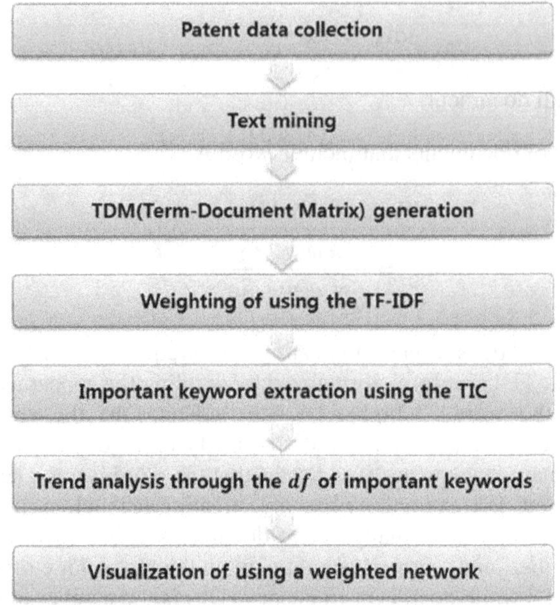

Fig. 1. Research methodology

First, we collect patent data of the target technology through a patent database such as Korea Intellectual Property Rights Information Service (KIPRIS) or WIPS ON [8, 9]. We then transform the collected data into analyzable data using a text mining technique. We remove stop words such as special letters, postpositions, and white space, and transform capital letters into small letters for all text. We derive a term–document matrix (TDM) to easily extract keywords using preprocessed data as variables. We then transform the frequency of the term in TDM into a weight using TF-IDF.

Table 1. Example of weighted TDM

Term-document	Document1	Document2	Document3	Document4
Term A	w_{11}	w_{12}	w_{13}	w_{14}
Term B	w_{21}	w_{22}	w_{23}	w_{24}
Term C	w_{31}	w_{32}	w_{33}	w_{34}
Term D	w_{41}	w_{42}	w_{43}	w_{44}

Most existing studies on deriving keywords through the application of TF-IDF have focused on TF-IDF weighting between keywords and documents. In these studies, keywords that represent each document are derived using the threshold of TF-IDF weighting determined in [10].

In this study, we analyze the problem of keyword determination by considering only the TF-IDF weighting, and propose a methodology to solve this problem. Before introducing the proposed methodology, we analyze the problem of extracting keywords by considering only the TF-IDF weighting using the following example.

Table 2. Example of weighted TDM 2

Term-document	Document1	Document2	Document3	Document4
Term A	0.5	0	0	0
Term B	0.3	0.3	0.4	0.1
Term C	0.08	0.05	0	0.1
Term D	0.01	0.03	0.7	0.05

Table 2 shows four patterns of weighted TDM. Term A appears only in Document 1, and the TF-IDF weighting of Term A is 0.5. If it is determined that the threshold of TF-IDF weighting is 0.25, term A can be considered a representative word of Document 1. The TF-IDF weight of term B exceeds 0.25 in Document 1, Document 2, and Document 3. Term B can therefore be considered as a core keyword of these three documents. Term B is a representative core keyword of many documents. For this reason, term B can be considered a principal and important term of these document sets. On the other hand, the TF-IDF weight of term C does not exceed 0.25 in any document. Therefore, we regard term C as a noise term. In the case of term D, it appears in all documents but its TF-IDF weight exceeds the threshold (0.25) in only Document 3. This pattern is the main issue of this study. As term D follows this pattern, it was selected as a keyword based on the TF-IDF threshold in existing studies. When we checked a term that has the same pattern as term D in the actual experimental data, we found it was actually not a keyword. We qualitatively analyzed documents that include this term. The results show that the term cannot explain the document. In actuality, term D is similar to a word that has the same pattern as term C, rather than the patterns of terms A and B. Term D is classified as noise because the pattern of term D is similar with the pattern of the noise terms. Term D tends to have a low TF-IDF weight in most documents, but it appears often enough to be classified as a core keyword in certain documents. However, term D is still meaningful in Document 3. The significance of term D was verified based on the TF-IDF weight in Document 3. Even if term D is not a keyword, it seemed like a keyword of Document 3 because term D had significance in document3. For instance, the term 'titanium' was derived as a keyword in the document, 'Titanium oxide coated carbon fiber and porous titanium oxide coated material composition,' which also includes 'coating,' such as in the pattern of term D. Although 'coating' is not a keyword in the document, it was determined to be a meaningful word along with 'titanium' because the two words are inter-related. These words may be helpful in explaining the technology of patent documents with a keyword using a data mining technique [11]. But it also confuses to select keyword in the big data. We therefore propose a new criterion for removing any confusion in the process of deriving keywords.

$$TIC_{(TF-IDF\ Criterion)} = \frac{\sum_1^N w_{kj}}{df_A} \tag{4}$$

In Eq. (4) above, TIC is the value used to divide the sum of w_{kj}, which signifies the importance of word A in a certain document, into df_A, which is the number of documents that include word A, and is used as a criterion for selecting a keyword. As shown in Figure 4, we select terms A and B as keywords when the threshold of TIC is 0.25. The TIC value distinguishes terms A and B from terms C and D more significantly for large amounts of data. Term A is the selected keyword of Document 1, and term B is the selected keyword of Document 1, Document 2, and Document 3, based on TIC. This shows that the technology represented by term B occupies much of the carbon fiber field. In addition, terms A and B can explain the technology of Document 1 together because they were derived as keywords in this document. We can also visually grasp a technology trend based on the results analyzed using a weighted network analysis.

6 Experiment

In this study, we analyzed the technology trend of carbon fiber using patent data. We collected 147 patents granted through a patent cooperation treaty (PCT) from 2008 to 2011 as experimental data. The collected patent data were restricted to representative documents to prevent an overlapping of text. We extracted the titles and abstracts from the collected patent data for use as analysis data.

We generated a text-document matrix from the data that were preprocessed using text mining. We then calculated the TIC score using the TF-IDF weight and document frequency. We then compared the TIC value with the maximum value of TF-IDF. When the threshold was 0.25, we derived the top-ten words that have a TIC value of less than the threshold and maximum TF-IDF value.

Table 3. Noises derived by TIC

Keyword	Document Frequency	TIC	Maximum of TF-IDF	Title of invention
Oxide	6	0.197713	0.576839	TITANIUM OXIDE COATED CARBON FIBER AND POROUS TITANIUM OXIDE COATED CARBON MATERIAL COMPOSITION
Sizing	8	0.20277	0.499945	CARBON FIBER
Copolymer	5	0.15286	0.493255	ACRYLONITRILE COPOLYMER AND METHOD FOR MANUFACTURING THE SAME AND ACRYLONITRILE COPOLYMER

Table 3. (*continued*)

Superior	3	0.214499	0.510428	APPARATUS AND PROCESS FOR PREPARING SUPERIOR CARBON FIBERS
Film	5	0.152525	0.517337	FINE CARBON FIBER DISPERSION FILM AND PROCESS FOR PRODUCTION OF THE FILM
Hot	3	0.235973	0.52969	POLYACRYLONITRILE FIBER MANUFACTURING METHOD AND CARBON FIBER MANUFACTURING METHOD
Element	6	0.236335	0.477384	CARBON FIBER AND CATALYST FOR CARBON FIBER PRODUCTION
Layered	3	0.233908	0.476721	LAYERED CARBON FIBER PRODUCT PREFORM AND PROCESSES FOR PRODUCING THES
Base	4	0.224731	0.472697	METHOD FOR CUTTING CARBON FIBER BASE
Content	10	0.115169	0.456205	CARBON FIBRE COMPOSTIONS COMPRISING LIGNIN DERIVATIVES

As Table 3 above shows, most of the derived words are not keywords of a patent. The word 'oxide' is involved with 'titanium,' which is a keyword in a document. The term 'sizing' does not indicate 'carbon fiber' as a keyword in a patent document. In the case of 'copolymer,' it has been mainly used for most manufacturing processes of carbon fiber. Therefore, it is not a keyword itself, but rather a word that appears along with the keyword 'Acrylonitrile' in a document. However, 'film' and 'layered' were found to be keywords when we qualitatively analyzed each document. When the TIC value of 'layered' was 0.233908, we could reduce any errors through an optimization of the threshold. However, when the TIC value of 'film' was 0.152525, an error occurred even though this value is relatively less than the threshold. The reason for this error can be explained in the next step.

Table 4. Keywords derived using TIC and the maximum value of TF-IDF

Keyword	Document frequency	TIC	Maximum of TF-IDF
Lignin	5	0.620363	1.026894
Films	3	0.497115	0.689526
Flexible	3	0.418967	0.606996
Softwood	2	0.842938	1.033279
Tow	2	0.756508	0.774959
Titanium	2	0.594135	0.774959
zinc	2	0.499805	0.929951
Cavity	2	0.49375	0.516639
Mandrel	2	0.465528	0.657541
Compositions	2	0.440663	0.516639
Mass	2	0.428991	0.522864
Kraft	2	0.421469	0.516639
Spheres	1	1.066618	1.066618
Prosthetic	1	0.959956	0.959956

After we removed any noise using TIC, We counted the number of documents that the represented keyword had selected using the TIC and maximum TF-IDF values. In addition, we defined the importance of the technology based on the number of documents.

We selected the top-fourteen words in order of document frequency. Five of the 147 patents collected had the keyword 'lignin.' The terms 'film' and 'flexible' were each found in three patents. The terms 'softwood,' 'tow,' 'titanium,' 'zinc,' 'cavity,' 'mandrel,' 'compositions,' 'mass,' and 'prosthetic' represented two patents each, and 'spheres' and 'prosthetic' were found in one patent each.

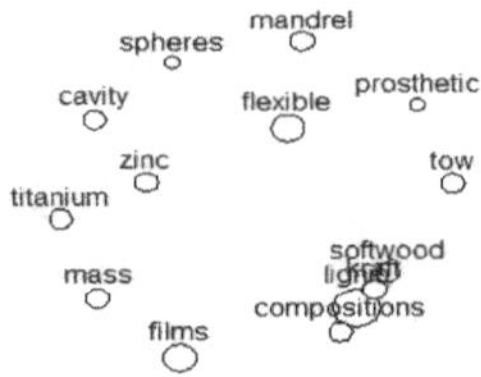

Fig. 2. Weighted network graph

Figure 2 shows a weighted graph indicating the importance of a technology visually. The size of the node represents the importance of each technology. In addition, inter-correlated words are linked together by an edge.

7 Conclusion

First, the keyword 'films' is synonymous with the term 'film,' which is classified as noise. The reason why 'film' was classified as noise is not due to the filtering of synonyms during the text mining process. It is therefore necessary to conduct further research on synonym filtering in text mining. Also we need to consider other languages for analyzing text [12]. This study proposed a method for selecting keywords quickly and accurately for large amounts of data. In addition, we visualized the keywords using a weighted network graph. This method provides more objective results than a qualitative method conducted by experts, and will be useful for eliminating noises in future studies on keyword determination.

A performance evaluation of TIC and a threshold optimization should be further studied. In addition, a future study will be to define a technology using the correlation between a keyword and words associated with the keyword.

Acknowledgement. This work was supported by the Brain Korea 21+ Project in 2013 This research was supported by the Basic Science Research Program through the National Research Foundation of Korea (NRF) funded by the Ministry of Education, Science, and Technology (No. 2010-0024163)

References

1. Korean Intellectual Property Office, Korea Invention Promotion Association.: Patent and information analysis (for researchers). Kyungsung Books, pp. 302–306 (2009)
2. Jun, S.H.: An efficient text mining for patent information analysis. In: Proceedings of KIIS Spring Conference, vol. 19(1), pp. 255–257 (2009)
3. Kim, J.S.: A knowledge Conversion Tool for Expert systems. International Journal of Fuzzy Logic and Intelligent Systems 11(1), 1–7 (2011)
4. Kang, B.Y., Kim, D.W.: A Muti-resolution Approach to restaurant Named Entity Recognition in Korean web. International Journal of Fuzzy Logic and In telligent Systems 12(4), 277–284 (2012)
5. Jung, C.W., Kim, J.J.: Analysis of trend in construction using textmining method. Journal of the Korean Digital Architecture Interior Association 12(2), 53–60 (2012)
6. Lee, S.J., Kim, H.J.: Keyword extraction from news corpus using modified TF-IDF. Society for e-business Studies 14(4), 59–73 (2009)
7. Myunghoi, H.: Introduction to social network analysis using R. Free Academy, 1–24 (2010)
8. Korea Intellectual Property Rights Information Service, `http://www.kipris.or.kr`
9. WIPS ON, `http://www.wipson.com`
10. Juanzi, L., Qi'na, F., Kuo, Z.: Keyword extraction based on TF-IDF for Chinese new document. Wuhan University Journal of Natural Sciences 12(5), 917–921 (2007)
11. Yoon, B.U., Park, Y.T.: A text-mining-based patent network: Analytical tool for high-technology trend. The Journal of High Technology Management Research 15(1), 37–50 (2004)
12. Jung, J.H., Kim, J.K., Lee, J.H.: Size-Independent Caption Extraction for Korean Captions with Edge Connected Components. International Journal of Fuzzy Logic and Intelligent Systems 12(4), 308–318 (2012)

Technology Analysis from Patent Data Using Latent Dirichlet Allocation

Gabjo Kim[1,3], Sangsung Park[2,3], and Dongsik Jang[1,3,*]

[1] Division of Industrial Management Engineering, Korea University
[2] Graduate School of Management of Technology, Korea University
[3] 1, 5-Ka, Anam-dong Sungbuk-ku, Seoul, 136-701 Korea

Abstract. This paper discusses how to apply latent Dirichlet allocation, a topic model, in a trend analysis methodology that exploits patent information. To accomplish this, text mining is used to convert unstructured patent documents into structured data. Next, the term frequency-inverse document frequency (tf-idf) value is used in the feature selection process. After the text preprocessing, the number of topics is decided using the perplexity value. In this study, we employed U.S. patent data on technology that reduces greenhouse gases. We extracted words from 50 relevant topics and showed that these topics are highly meaningful in explaining trends per period.

Keywords: latent Dirchlet allocation, topic model, text mining, tf-idf.

1 Introduction

Recently, an efficient technique has become necessary to rapidly analyze trends in the field of science technology. Further, this efficient technique needs to facilitate and cope well with research and development (R&D) for companies' new products that are motivated by products of competitors or leading companies. The technique begins by analyzing patents, which hold important technology and intellectual property information. By analyzing patents, companies and countries are able to monitor technology trends and to evaluate the technological level and strategies of their competitors.

Intellectual property information is a useful tool when managing an R&D strategy because it is the best way to monitor competitors' new technology and technology trends. This information is also a better reflection of market dominance than papers and journals. Leading global companies know that intellectual property information is the one of the best tools to use to manage the R&D function of the company. However, many companies still use intellectual property as a simple form of R&D, but if they wait until after the results of the research to analyze patent applications, this is equivalent to navigating after arriving at a destination. Thus, in this paper, we suggest a technology trend analysis methodology that uses patents in intellectual property, including technology information.

* Corresponding author.

K.M. Lee, S.-J. Park, J.-H. Lee (eds.), *Soft Computing in Big Data Processing*,
Advances in Intelligent Systems and Computing 271,
DOI: 10.1007/978-3-319-05527-5_8,

Some technology trend studies have been done by applying patents. Most of these studies use the time lag of a patent document, citation information, and family patent information. Furthermore, as the abstracts and technology information are primarily in the form of unstructured data, these have to be converted into structured data to analyze keywords using text mining [1]. However, these methods are limited in that the results of the analysis do not yield much about the technology information in the patent document.

In this study, we suggest a different approach. Here, we use a topic model, introduced by Blei, which applies the concept of a generative probabilistic model to patents to grasp topic trends over time. The topic model is usually used for machine translation of a document and search engines. In our case, a topic is expressed in terms of a list of words and people can define a topic using words. Finally, the topic model can analyze the latent pattern of documents by using extracted topics. In this way, our approach resolves the limitation of existing approaches that find it difficult to extract detailed industry technology information.

2 Literature Review

A patent document includes substantial information about research results [2]. Generally, a patent application is submitted prior to the R&D report or other papers. Furthermore, in terms of future-oriented advanced technology, patent analysis data are the most basic data for judging trends in the relevant field.

Recent studies have suggested forecasting a vacant area of a given technology field by using patent analyses. Of these, some studies cluster patent documents and identify the lower number of clustered patent documents as promising future technology. Jun et al. proposed a methodology for discovering vacant technology by using K-medoids clustering based on support vector clustering and a matrix map. They collected U.S., Europe, and China patent data in the management of technology (MOT) field to cluster patent documents and defined this as the vacant technology [3]. Yoon et al. have suggested a patent map based on a self-organizing feature map (SOFM) after text mining. They analyzed vacant technology, patent disputes, and technology portfolios using patent maps [4].

Previous studies that apply patent information to research into vacant technology are limited by a lack of objectivity in their qualitative techniques, including the subjective measures of the researchers. In addition, most of these studies do not consider time series information in patent documents that contain rapidly changing trends. Therefore, in this study, we convert patent documents into structured data using text mining and apply a topic model to overcome the limitations in existing research to grasp trends objectively.

The objective of topic model is to extract potentially useful information from documents [5]. Many topic model studies have also been suggested. Blei et al. conducted a study to examine the time course in the topic. They validated their model to OCR'ed archives of the journal Science from 1880 to 2000 [6]. In addition, Griffiths et al. conducted a study using a topic model by extracting scientific topics to examine scientific trends. They analyzed the 1991 to 2001 corpus of the Proceedings of the National Academy of Sciences of the United States of America (PNAS) journal abstracts and determined the hottest and coldest topics during this period [7].

To the best of our knowledge, no existing studies apply a topic model to researching trend analyses. In addition, since the advantage of the topic model is its ability to extract the topic of each document, this is an effective way to analyze patent documents. We provide an easy way to understand the trend analysis methodology that applies the topic model to patent documents, including technology information.

3 Methodology

3.1 Data Preprocessing

As patent data is typically unstructured data, text mining as a text preprocessing step is needed to analyze the data using statistics and machine learning. Text mining is a technology that enables us to obtain meaningful information from unstructured text data. Text mining techniques can help extract meaningful data from huge amounts of information, grasp connections to other information, and discover the category that best describes the text. To achieve this, we eliminated stop words and numbers, converted upper case to lower case, and eliminated white space within sentences. To then analyze a patent document, we converted it into a document-term matrix (DTM) structure.

3.2 The Tf-Idf

After converting patent data into the DTM, the process of feature selection is required. Feature selection is an important process to evaluate subset [8][9]. There are several methods on feature selection. In this study, we adopted the term frequency-inverse document frequency (tf-idf). The tf-idf is a method used to assign weights to important terms in a document. The documents represented in the tf-idf weights are able to determine those documents that are the most similar to a given query and make it simple to cluster similar documents. A big tf-idf value makes it easiest to determine the subject of the document and this weight is a useful index with which to extract the core keywords. With given documents, term frequency (TF) simply refers to how many times a term appears in the document. In this case, if the document is long, the term frequency in the document will be larger, regardless of importance. The TF is defined by

$$tf_{i,j} = \frac{n_{i,j}}{\sum_k n_{k,j}} \tag{1}$$

where $n_{i,j}$ is the frequency of term t_i in document d_j, and the denominator is the frequency of the total number of terms in document d_j.

The inverse document frequency (IDF) is a value that represents the importance of relevant terms. It is obtained by dividing the total number of documents by the number of documents containing the relevant term, and then taking the logarithm of the quotient. The IDF is represented by

$$idf_i = log\frac{|D|}{|\{d_j : t_i \in d_j\}|} \quad (2)$$

where $|D|$ is total number of documents in the corpus, and $|\{d_j : t_i \in d_j\}|$ represents the number of the documents contain term t_i.

The tf-idf weight is calculated by multiplying the TF by the IDF. Therefore, this weight allows us to omit terms that have a low frequency, as well as those that occur in many documents. In this study, we considered tf-idf values of 0.1 and over as important terms and applied them to our analysis.

3.3 Latent Dirichlet Allocation (LDA)

Topic models based on latent Dirichlet allocation (LDA) are generative probabilistic models. In these models, the parameter and the probability distribution randomly generate observable data. With a modeling document, if we know the topic distribution of the document and the generating probability of terms for each topic, it is possible to calculate the generating probability of the document. LDA consists of three steps [10]:

Step 1 : Choose N~Poisson(ξ).
Step 2 : Choose θ~Dirichlet(α).
Step 3 : For each of the words w_n:
Choose topic z_n~Multinomial(θ).
Choose a word w_n from a multinomial probability distribution conditioned on the topic z_n : $p(w_n|z_n,\beta)$.

In the document, we first choose the number of words in the documents, N, which is drawn from a Poisson distribution (Step 1). Second, the topic distribution per document is drawn from the Dirichlet distribution and given the Dirichlet parameter, , which is a K-vector with components $\alpha_i > 0$, as shown in Eq. (3).

$$p(\theta|\alpha) = \frac{\Gamma(\sum_{i=1}^{K}\alpha_i)}{\prod_{i=1}^{K}\Gamma(\alpha_i)}\theta_1^{\alpha_1-1}\cdots\theta_K^{\alpha_K-1} \quad (3)$$

Given the parameters α and β, the joint distribution of a topic mixture, θ, the topics, z, along with their proportions, and the words, w, with given priors is represented in Eq. (4).

$$p(\theta,z,w|\alpha,\beta) = p(\theta|\alpha)\prod_{n=1}^{N}p(z_n|\theta)p(w_n|z_n,\beta) \quad (4)$$

Finally, the probability of a corpus can be calculated by taking the product of the marginal probabilities of the individual documents, as shown in Eq. (5) and (6).

$$p(w|\alpha,\beta) = \int p(\theta|\alpha)\left(\prod_{n=1}^{N}\sum_{z_n}p(z_n|\theta)p(w_n|z_n,\beta)\right)d\theta \quad (5)$$

$$p(D|\alpha,\beta) = \prod_{d=1}^{M} \int p(\theta_d|\alpha)\left(\prod_{n=1}^{N_d} \sum_{z_{dn}} p(z_{dn}|\theta_d)p(w_{dn}|z_{dn},\beta)\right)d\theta_d \quad (6)$$

In this paper, we used variational expectation maximization (VEM) algorithm to infer hidden variables [10]. The LDA can simply be denoted as a graphical model in plate notation, as shown in Figure 1.

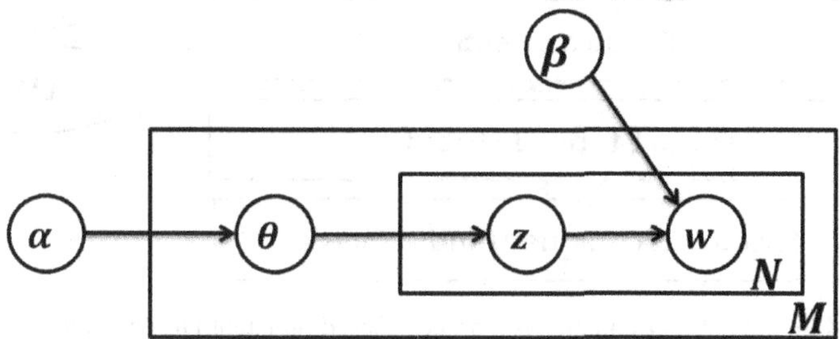

Fig. 1. A graphical model of the latent Dirichlet allocation

The nodes denote random variables and the edges are the dependencies between the nodes. The rectangles refer to replication, with inner rectangles representing words and outer rectangles representing documents.

The LDA model has an important assumption that the words in the documents are not arrayed in an orderly way (the bag of words assumption) [11]. Therefore, the order of the words is ignored in the LDA model, although the word-order information might contain important clues as to the content of a document

3.4 Determining the Number of Topics

When modeling with a topic model, determining the number of topics needs to be fixed a-priori. Generally, as the number of topics is unknown, several different numbers of topics are fitted into the model and the optimal number of topics is determined in a data-driven way. In addition, it is possible to determine the number of topics by dividing the data into training and test data [12]. Therefore, we split the data into training and test data sets to validate by tenfold cross validation, and then apply perplexity to determine the optimal number of topics. The perplexity is represented by Eq. (7).

$$\text{perplexity} = \exp\left\{-\frac{\sum_{d=1}^{M} \log p(w_d)}{\sum_{d=1}^{M} N_d}\right\} \quad (7)$$

where w_d is the total number of words and N_d is the total number of words in document d respectively. When the perplexity value of perplexity is lower, the optimal topic is set to [7].

Figure 2 illustrates a broad overview of the trend analysis process based on LDA.

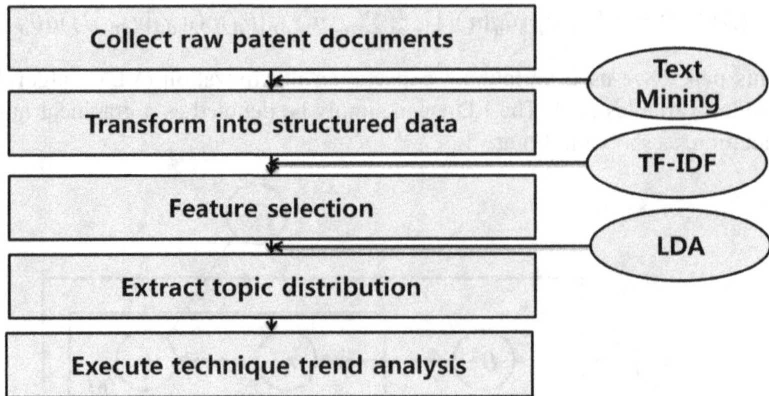

Fig. 2. Overall trend analysis process based on latent Dirichlet allocation

4 Experimental Results

In our results, we collected U.S. patent data on technology that reduces greenhouse gases from 2000 to 2012, as shown in Figure 3. The applications for and the published patent data on this technology was highest in 2003, and lowest in 2012 because these patents were still in the process of judgment and not yet published.

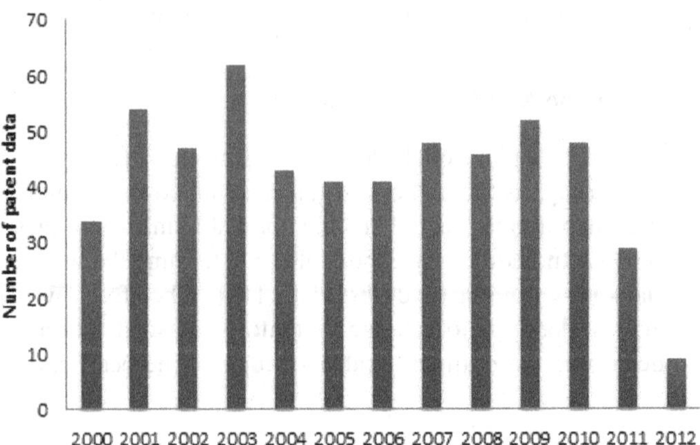

Fig. 3. U.S. patent data for experiment from 2000 to 2012

We carried out text preprocessing by converting capital letters to lowercase letters and removing Arabic numerals and spaces between sentences to make the analysis easier. And mentioned earlier, we used terms with a tf-idf value of 0.1 and over for feature selection. Figure 4 shows the results of using perplexity as a measure to select the optimal topics in the process of extracting topics.

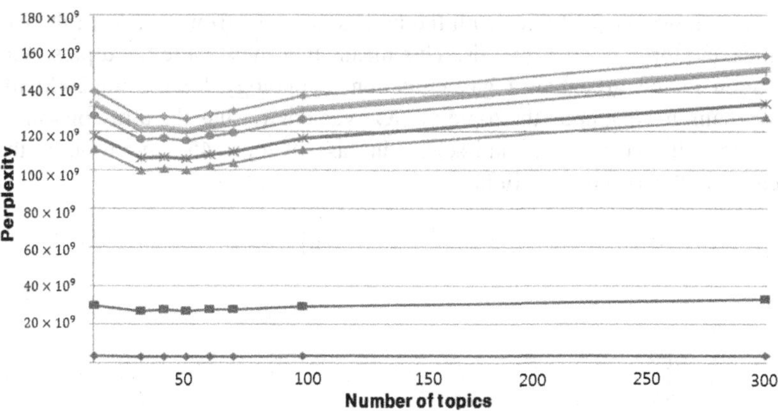

Fig. 4. Perplexity of the test data for the LDA

According to the minimum perplexity values, we determined that the optimal number of topics for the model would be 50.

Table 1. The 12 most important topics containing five important words

Topic 45	**Topic 16**	**Topic 11**	**Topic 48**	**Topic 14**	**Topic 36**
NOx	Coat	PFC	Swing	Waste	Solar
Regenerator	Plasma	Clean	Purity	Sulfid	Nitric
Silicon	Layer	Install	Abatement	Ferment	Off-gas
Passage	Silica	Inject	Cdo	Salts	Biomass
Sensor	Thermophil	Etcher	Biology	Pond	Solar-driven
Topic 4	**Topic 46**	**Topic 37**	**Topic 26**	**Topic 35**	**Topic 32**
Impur	Fluorine	Polym	Layer	Feedstock	Substrate
Agent	Choloride	Matrix	Input	Gasifty	Tetrafluoropropen
Pore	Fluoride	Film	Region	Units	Thereafter
Value	Trifluoroethan	Molecular	Bacteria	Operation	Shape
Underground	Difluoromethhan	Micropor	Appi	Farm	Product

Table 1 shows the 12 topics with the highest topic distribution, θ, of the extracted 50 topics and top 5 relevant words. This means that these were the topics mentioned most often in the greenhouse gas reduction technology field from 2000 to 2012. Among them, for example, the topic of NOx gas control has the array of words NOx, regenerator, silicon, passage, and sensor in case of topic 45. This topic is the most referenced in this technology field.

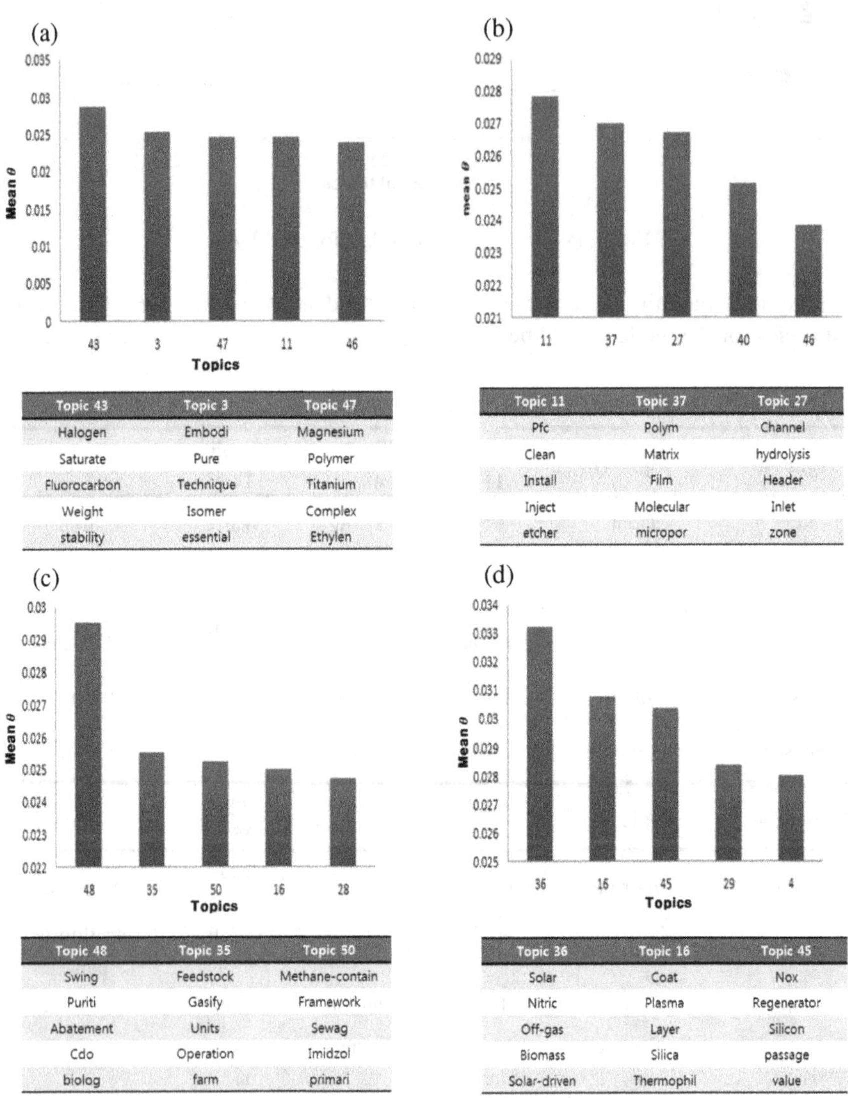

Topic 43	Topic 3	Topic 47
Halogen	Embodi	Magnesium
Saturate	Pure	Polymer
Fluorocarbon	Technique	Titanium
Weight	Isomer	Complex
stability	essential	Ethylen

Topic 11	Topic 37	Topic 27
Pfc	Polym	Channel
Clean	Matrix	hydrolysis
Install	Film	Header
Inject	Molecular	Inlet
etcher	micropor	zone

Topic 48	Topic 35	Topic 50
Swing	Feedstock	Methane-contain
Puriti	Gasify	Framework
Abatement	Units	Sewag
Cdo	Operation	Imidzol
biolog	farm	primari

Topic 36	Topic 16	Topic 45
Solar	Coat	Nox
Nitric	Plasma	Regenerator
Off-gas	Layer	Silicon
Biomass	Silica	passage
Solar-driven	Thermophil	value

Fig. 5. Topics of top 5 mean θ for (a) early 2000s, (b) mid 2000s, (c) late 2000s, and (d) 2010s

Figure 5 indicates the five top-ranked θ values of topics for the early 2000s, mid-2000s, late 2000s, and 2010s. Topics about methods of producing fluorocarbons and topics about purifiers of mixtures featured strongly in the early 2000s. Topics about purifier of PFCs and improvements in the efficiency of mixture matrix membranes were popular in the mid-2000s. Topics about impurity abatement using CDO chambers and disassembling nitrogenous compounds using solar-driven techniques showed strongly in the late 2000s and 2010s, respectively. This actually demonstrates that there is active support and development by the U.S. government for its goal of developing and diffusing climate change response technology through its Climate Change Technology Initiative (CCTI) [13].

5 Conclusion

In this paper, we presented LDA, a generative model for documents containing words, and applied this model to patent documents to determine trends. To do this, we first converted the unstructured abstracts into structured data. We then used the tf-idf values with a weight of 0.1 and over as analyzable words. We determined the optimal number of topics using perplexity and understood the topic trend using the most common words in that topic. This is the first attempt at a technique for trend analysis by extracting information from patent topics.

We experimented with the proposed method using data on technology that reduces greenhouse gases. As a result, we found that producing fluorocarbons in the early 2000s, an apparatus for purifying PFCs in the mid-2000s, impurities abatement using CDO chambers in the late 2000s, and nitrogen compounds using solar heat in the 2010s were the featured topics in each period. It was possible to determine the technical trends in technology to reduce greenhouse gases by looking at the topics in a dynamic period and the words within each topic in the patent documents.

The advantage of this study is that we recognize a specific trend using topics within a period, which has not been attempted before. However, there are limits to inferring the meaning of a topic from the words in the topic unless those who do so are experts in the field. Moreover, we can expect better experimental results by indicating more visually whether there is a biased topic distribution based on time. Applying clustering to these extracted topics to study vacant technology might be more meaningful in future research. Furthermore, we expect further studies to apply this approach to technology fields other than reducing greenhouse gases and to forecast future trends using a time series analysis.

Acknowledgement. This work was supported by the Brain Korea 21 Plus Project in 2013. This research was supported in part by the Basic Science Research Program through the National Research Foundation of Korea (NRF) funded by the Ministry of Education, Science, and Technology (No. 2010-0024163).

References

[1] Lee, S.J., Yoon, B.Y., Park, Y.T.: An approach to discovering new technology opportunities: Keyword-based patent map approach. Technovation 29, 483–484 (2009)

[2] Tseng, Y.H., Lin, C.J., Lin, Y.I.: Text mining techniques for patent analysis. Information Processing & Management 43, 1216–1247 (2007)

[3] Jun, S.H., Park, S.S., Jang, D.S.: Technology forecasting using matrix map and patent clustering. Industrial Management & Data Systems 112(5), 786–807 (2012)
[4] Yoon, B.U., Yoon, C.B., Park, Y.T.: On the development and application of a self-organizing feature map-based patent map. R&D Management 32(4), 291–300 (2002)
[5] Noh, T.G., Park, S.B., Lee, S.J.: A Semantic Representation Based-on Term Co-occurrence Network and Graph Kernel. International Journal of Fuzzy Logic and Intelligent Systems 11(4) (2011)
[6] Blei, D.M., Lafferty, J.D.: Dynamic Topic Models. In: 23rd International Conference on Machine Learning, Pittsburgh, PA (2006)
[7] Griffiths, T.L., Steyvers, M.: Finding scientific topics. Proceedings of the National Academy of Sciences of the United States of America 101, 5228–5235 (2004)
[8] Uhm, D., Jun, S., Lee, S.J.
[9] Cho, J.H., Lee, D.J., Park, J.I., Chun, M.G.: Hybrid Feature Selection Using Genetic Algorithm and Information Theory. International Journal of Fuzzy Logic and Intelligent Systems 13(1) (2013)
[10] Blei, D.V., Ng, A.Y., Jordan, M.I.: Latent Dirichlet Allocation. Journal of Machine Learning Research 3, 993–1022 (2003)
[11] Steyvers, M., Griffiths, T.: Probabilistic topic models
[12] Grun, B., Hornik, K.: topicmodels: An R Package for Fitting Topic Models. Journal of Statistical Software 40(13) (2011)
[13] Simpson, M.M.: Climate Change Technology Initiative (CCTI): Research, Technology, and Related Program. CRS Report for Congress (2001)

A Novel Method for Technology Forecasting Based on Patent Documents

Joonhyuck Lee[1,3], Gabjo Kim[1,3], Dongsik Jang[1,3], and Sangsung Park[2,3,*]

[1] Division of Industrial Management Engineering, Korea University
[2] Graduate School of Management of Technology, Korea University
[3] 1, 5-ga, Anam-dong, Seongbuk-gu, Seoul, 136-701 Korea

Abstract. There have been many recent studies on forecasting emerging and vacant technologies. Most of them depend on a qualitative analysis such as Delphi. However, a qualitative analysis consumes too much time and money. To resolve this problem, we propose a quantitative emerging technology forecasting model. In this model, patent data are applied because they include concrete technology information. To apply patent data for a quantitative analysis, we derive a Patent–Keyword matrix using text mining. A principal component analysis is conducted on the Patent–Keyword matrix to reduce its dimensionality and derive a Patent–Principal Component matrix. The patents are also grouped together based on their technology similarities using the K-medoids algorithm. The emerging technology is then determined by considering the patent information of each cluster. In this study, we construct the proposed emerging technology forecasting model using patent data related to IEEE 802.11g and verify its performance.

Keywords: Emerging technology, Patent, Technology Forecasting.

1 Introduction

There are many recent studies on forecasting emerging and vacant technologies [1-3]. Most of them depend on a qualitative analysis such as Delphi. However, a qualitative analysis takes too much time and money. To solve this problem, we propose a quantitative emerging technology forecasting model. In this model, patent data are applied because such data include concrete information of a technology. To apply patent data for a quantitative analysis, we derive a Patent–Keyword matrix using text mining. A principal component analysis is conducted on the Patent–Keyword matrix to reduce its dimensionality and derive a Patent–Principal Component matrix. The patents are also grouped together based on their technology similarities using the K-medoids algorithm. Next, the emerging technology is determined by considering the patent information of each cluster. In this study, we construct the proposed emerging technology forecasting model using patent data related to IEEE 802.11g and verify its performance.

---

[*] Corresponding author.

K.M. Lee, S.-J. Park, J.-H. Lee (eds.), *Soft Computing in Big Data Processing*, Advances in Intelligent Systems and Computing 271,
DOI: 10.1007/978-3-319-05527-5_9,

2 Literature Review

2.1 Patent Information

Patent information is a representative intangible asset and is standardized into a specific form. In addition, we can read about recent technology trends by analyzing patent data through an early disclosure system [4]. For this reason, patent information has been frequently used for quantitative analyses [5]. There have been many efforts to develop new patent indicators by combining several types of individual patent information and to try to use more patent indicators for a quantitative technology analysis [6]. The most basic patent information is the numbers of patent applications and patent citations. The number of patent applications can be considered an indicator showing the quantitative technology innovation activity of certain applicants such as research laboratories and companies. However, the technological completeness and economic value of different patents vary considerably. For this reason, there are limitations on evaluating technology innovation activities using only the number of patent applications. For this reason, patent citation information has been frequently applied as an indicator showing technological completeness and the economic value of the patent [7]. Additionally, patent citation information is used for deriving other patent indicators such as the cites per patent (CPP), patent impact index (PII), current impact index (CII), and technology impact index (TII) [2].

2.2 Technology Forecasting

Most existing researches on the prediction of emerging technologies have depended upon a qualitative analysis such as a survey or expert judgment. Recently, some studies on forecasting emerging technologies through quantitative methods have been conducted.

Jun applied patent information to a quantitative analysis for determining the vacant technology of intelligent systems [1]. In this study, a preprocessing method for applying patent documents to a quantitative analysis was conducted using a text mining method. In addition, the vacant technology of an intelligent system was derived from the clustering of preprocessed patent information.

Kim conducted a study to determine a vacant OLED technology [2]. In this study, a preprocessing method for applying patent documents to a quantitative analysis was conducted using a text mining method. In addition, the optimal number of clusters was derived using a Bayesian information criterion (BIC). The CLARA algorithm was applied in this study for deriving a vacant technology owing to its good performance on the clustering of large entities. Kim determined the cluster with the smallest size for a vacant technology of the OLED industry.

These two studies derived a vacant technology by considering only the size of each cluster. For this reason, they are at risk of selecting a worthless technology cluster as an emerging technology cluster because most clusters of worthless technologies have a small size. In addition, they are at risk of selecting technology clusters with a small size but that no longer require development as an emerging technology cluster.

To avoid these risks, in this study, patent information such as the patent application date, patent citation, and PFS is considered for the prediction of emerging technologies.

3 Emerging Technology Forecasting Model

In this study, we propose a model for forecasting an emerging technology by applying patent information. Fig. 1 shows a summary of the proposed model.

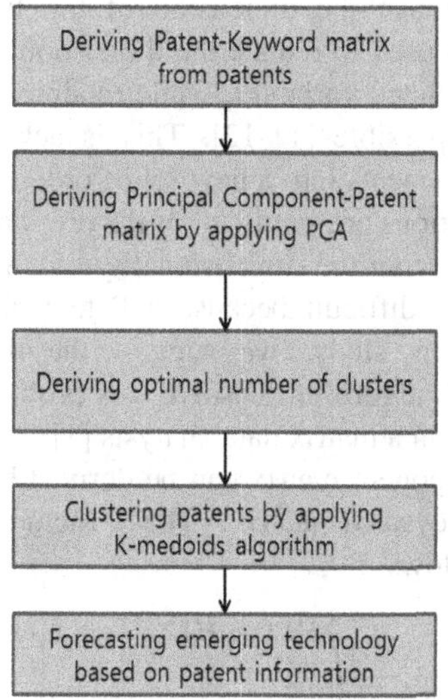

Fig. 1. Summary of the proposed model

3.1 Deriving a Patent–Keyword Matrix

A preprocessing method for retrieved patent documents is required for applying a patent document to a quantitative analysis. Retrieved patent documents are converted into the appropriate data form for data mining during the preprocessing stage. In this study, the TM Package of the R project is applied for converting patent documents into a Patent–Keyword matrix.

The structure of a Patent–Keyword matrix is as follows.

$$X_{n\times k} = \begin{pmatrix} x_{11} & x_{12} \cdots & x_{1k} \\ \vdots & \ddots & \vdots \\ x_{n1} & x_{n2} \cdots & x_{nk} \end{pmatrix}$$

In the above matrix, n indicates the number of patents, and k is the number of keywords. In addition x_{nk} represents the number of incidences of the k th keyword of the n th patent document. Generally, the number of keywords in a patent document is much greater than the number of patents used in an analysis. For this reason, the matrix above usually has much bigger columns than rows. In this case, the efficiency of the data analysis declines dramatically owing to the 'Dimensionality Curse' [8-10]. In this study, we suggest a method to solve this problem by conducting a dimension reduction on the Keyword–Patent matrix.

3.2 Deriving a Principal Component–Patent Matrix by Applying PCA

A principal component analysis (PCA) is a representative dimension reduction method, and we try to reduce the dimension of the Patent–Keyword matrix by applying a PCA. A PCA is used to reduce the dimensionality of a dataset consisting of a large number of interrelated variables, while retaining as much of the variations present in the dataset as possible [11-12]. This is achieved by transforming the uncorrelated principal components into a new set of ordered variables so that the first few retain most of the variations present in all of the original variables [11].

As described above, reducing the dimensionality of a Patent–Keyword matrix by applying a PCA directly is difficult because a Patent–Keyword matrix has larger columns than rows. In this study, we suggest the use of the singular value decomposition (SVD)–PCA algorithm to solve this problem. SVD is a method that solves calculative problems in a matrix data analysis [1].

A Patent–Principal Component matrix can be derived by applying the SVD–PCA algorithm on a Patent–Keyword matrix. The structure of the Patent–Principal Component matrix is as follows.

$$P_{n\times f} = \begin{pmatrix} p_{11} & p_{12} & \cdots & p_{1f} \\ \vdots & & \ddots & \vdots \\ p_{n1} & p_{n2} & \cdots & p_{nf} \end{pmatrix}$$

In the matrix above, n indicates the number of patents, and f is the number of principal components. These principal components were derived from the keywords using a PCA. In addition, p_{nf} shows the principal component score of the f th principal component of the n th patent document.

3.3 Clustering Patent Document by Applying K-medoids Algorithm

In this study, we apply a clustering algorithm to group patents together based on their technology similarities. A typical clustering method is hierarchical clustering. Hierarchical clustering is the grouping of objects in the same cluster that are more similar to each other than to objects in other clusters. Typical hierarchical clustering algorithms include the K-means algorithm and K-medoids algorithm. The K-medoids algorithm usually shows a better performance than the K-means algorithm when the dataset used for the clustering has many outliers and a lot of noise [13]. Generally, a lot of outliers and noise can be found in patent documents. For this reason, we try to apply the K-medoids algorithm for patent-document grouping. To apply the K-medoids algorithm to clustering, the appropriate number of clusters should be determined. There are many methods for deriving the optimal number of clusters, including the BIC, Akaike information criterion (AIC), and Silhouette width.

In this study, we apply the Silhouette width for deriving the optimal number of clusters because, using this method, more mathematical information such as the standard deviation and mode can be considered [14]. The Silhouette width shows how appropriate the number of clusters is to a dataset. The Silhouette width has a value between -1 and 1. A larger Silhouette width indicates a more optimal number of clusters.

3.4 Forecasting Emerging Technology Using Patent Information

In this study, we try to predict an emerging technology by considering the PFS, patent citation information, and application date. A patent family is a set of patents taken in various countries to protect a single invention [15]. Protecting an invention through a patent and applying for a patent in a foreign country is generally costly. For this reason, a bigger patent family size implies that the value of the patent is the much higher. The number of patent citations indicates the number of times a patent has been referenced by another patent since its initial disclosure. For this reason, we can think of a patent that has been referenced many times by other patents as a valuable and important one. On the other hand, we can think of clustered patents that have been referenced many times by other patents as having been fully developed already.

The patent application date is also disclosed when the patent is published. We can think of a cluster of older developed patents to be a sufficiently developed technology cluster.

For this reason, a patent cluster that has a relatively large PFS, a relatively small citation rate, and a relatively short amount of time since its development can be considered an emerging technology cluster. In this study, we propose a technology emerging index based on patent information. An equation of a technology emerging index is as follows.

$$\text{Technology Emerging Index} = \frac{n_k}{c_k \times y_k} \quad (1)$$

n_k= Average PFS of patents in cluster K
c_k= Average number of referenced patents in cluster K
y_k= Average number of years since the patent filing date in cluster K

4 Constructing and Verifying the Emerging Technology Model

In this section, we try to construct the proposed emerging technology forecasting (ETF) model and validate it using patent data related to 802.11g technology.

The reason for selecting 802.11g for this analysis is that this technology has been widely adopted in many IT devices. In addition, there have been many published patents related to 802.11g since its development, and it therefore makes it easy to construct and validate the proposed ETF model.

IEEE 802.11g is an information technology standard for wireless LAN. The 802.11g standard has been rapidly commercialized since its initial development in 2003. 802.11g works in the 2.4 GHz band, and has a faster data rate than the former 802.11b standard, with which it maintains compatibility. Owing to its fast data rate and compatibility with 802.11b, 802.11g has been widely adopted in many IT devices.

In this study, patent data related to IEEE802.11g was retrieved from the United States Patents and Trademark Office (USPTO) for analysis. Since the first patent related to IEEE802.11g was filed in 2002, a total of 271 patents have been published. Fig. 2 shows the changes in the number of filed patents related to IEEE802.11G by year.

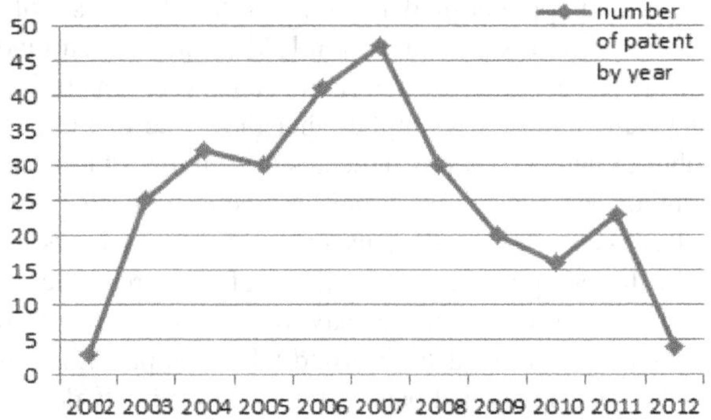

Fig. 2. Number of patents filed per year

We constructed a model for predicting an emerging technology using patent information and a clustering algorithm. To construct the model, we created two datasets: training datasets and test datasets. We constructed an ETF model using training patent data filed prior to 2008. In addition, we validated the constructed ETF model by applying patent data that had been filed between 2008 and 2012. Table 1 shows the training and test datasets.

Table 1. Summary of the training and test data

	year	number of patents
Training data	2002–2008	208
Test data	2009–2012	63

In this study, we applied a program package of the R project ('tm' and 'cluster') for the preprocessing, PCA, and clustering process. We used exemplary claims from 208 training patent data for deriving a Patent–Keyword matrix that can be applied to generate the ETF model.

First, we used a preprocessing method to train the data into an appropriate data form for a quantitative analysis. After extracting the keywords from the patent document using text mining, we obtained a Patent–Keyword matrix. The dimensions of this matrix are 208 × 2953. The number of columns of the obtained Patent–Keyword matrix is much greater than the number of rows. For this reason, we were unable to use PCA directly to obtain the Patent–Principal Component matrix. To solve this problem, we used the SVD–PCA to obtain this matrix. Using the SVD–PCA, we obtained a total of 208 principal components. Among the 208 principal components, we selected the top 103 principal ones for obtaining the Patent–Principal Component matrix. The reason for selecting the top 103 principal components is that their cumulative percentage of variance exceeds 95%.

Before clustering the patents using the Patent–Principal Component matrix, we applied the Silhouette width to determine the number of clusters (k) for K-medoids clustering. Fig. 3 shows the changes in Silhouette width according to the number of clusters (k). We selected the optimal number of clusters from the result of the Silhouette width. We determined that the optimal number of clusters is three.

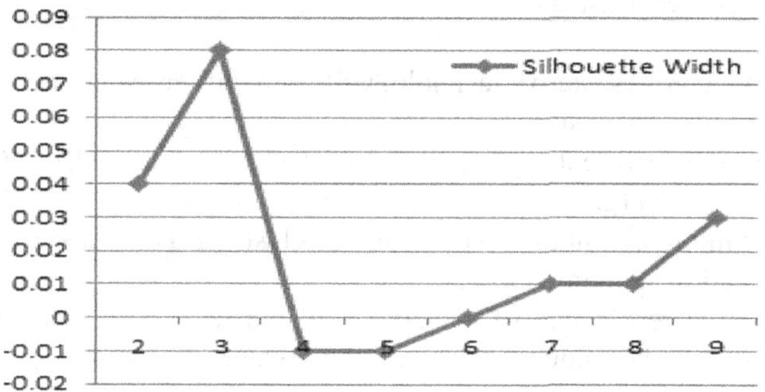

Fig. 3. Silhouette width according to the number of clusters

We conducted clustering for the Patent–Principal Component matrix using the K-medoids algorithm based on the optimal number of clusters determined.

Table 2 shows the K-medoids clustering results, including the top ten ranked keywords, average number of references, average PFS, average number of years since the filing date, and average technology emerging index score of each cluster.

Table 2. Result of K-medoids clustering

Cluster	Top 10 keywords in cluster	Ratio(%)	average number of references	PFS	average amount of time after filing date	TEI
1	signal, network, plurality, communication, comprising, data, packet, wireless, method, channel	65.87	7.26	17.146	7.58	0.312
2	configured, device, access, apparatus, module, communication, transmit, wireless, packets, transmit	9.13	2.894	9.36	6.79	0.476
3	control, envelope, frequency, output, plurality, power, signal, constant, phase, comprising	25	20.13	37.673	6.99	0.268

We defined the representative technology of each cluster using the most frequently occurring terms of each cluster.

We defined the representative technology of cluster 1 as a data communication method applying 802.11g.

We defined the representative technology of cluster 2 as a transceiving device for data communication applying 802.11g.

We defined the representative technology of cluster 3 as a technology that can control and regulate the frequency, power, and voltage changes of devices that apply 802.11g.

As shown in Fig. 2, the average amount of time after the filing date of patents included in cluster 1 is relatively longer than in the other clusters. In addition, the average number of referenced patents is higher, and many similar patents were already filed. For this reason, we determined the representative technology of cluster 1 as a sufficiently developed technology.

Cluster 3 has relatively high number of referenced patents and a large PFS in comparison with ratio of patents shown in Table 2. However, the technology emerging index of cluster 3 is lower than in the other clusters, and the average amount of time since the filing date is relatively long. For this reason, we determined the representative technology of cluster 3 as being an important but sufficiently developed one.

Cluster 2 has a higher technology emerging index than the other clusters, and the average amount of time since the filing date is shorter. For this reason, we determined the representative technology of cluster 2 to be an emerging technology.

We validated the performance of the ETF model using the test dataset. Table 3 shows the number of patents related to each of the three clusters.

Table 3. Number of patents of clusters 1, 2, and 3 in the test data

Cluster	number of patents	ratio
1	35	55.56%
2	25	39.68%
3	3	4.76%

As shown in Table 3, the ratio of patents of clusters 1 and 3 declined in comparison with the period between 2002 and 2008. However, the ratio of patents of cluster 2 was highly increased.

These results support the ability of the proposed model to properly predict emerging technologies.

5 Conclusion

Studies on forecasting emerging and vacant technologies quantitatively through the application of patent information have recently been conducted. However, most of these studies did not consider various types of patent information such as the citation information and filing date. Because of this limitation, there is a risk of selecting a worthless technology cluster as an emerging technology cluster. In addition, there is a risk of selecting small technology clusters that no longer require development as an emerging technology cluster. To solve this problem, we proposed an emerging technology forecasting model using various types of patent information such as the citation information, filing date, and patent family size. We constructed the proposed ETF model by applying patent information related to IEEE802.11g. The results of the constructed ETF model provide technology related to transceiving devices for data communication applying 802.11g as an emerging technology. Additionally, we verified the performance of the ETF model using a test dataset. Based on the results of this study, we can see that predicting an emerging technology by considering various types of patent information is more efficient than predicting an emerging technology by considering only the number of patents in a technology cluster.

In this study, we also proposed a technology emerging index using patent information, and verified this index empirically through a test dataset.

To improve the performance of this promising technology forecasting model, the development of more precise indicators to indicate a vacant or emerging technology is needed. In addition, a more precise technology forecasting will be made possible by constructing a technology forecasting model using both patents and major academic journals that include advanced technology information.

Acknowledgement. This work was supported by the Brain Korea 21+ Project in 2013. This research was supported by the Basic Science Research Program through the National Research Foundation of Korea (NRF) funded by the Ministry of Education, Science, and Technology (No. 2010-0024163)

References

[1] Jun, S.H., Park, S.S., Jang, D.S.: Technology forecasting using matrix map and patent clustering. Industrial Management & Data Systems 112(5), 786–807 (2012)

[2] Kim, Y.S., Park, S.S., Jang, D.S.: Patent data analysis using CLARA algorithm: OLED Technology. The Journal of Korean Institute of Information Technology 10(6), 161–170 (2012)

[3] Lee, S., Yoon, B., Park, Y.: An approach to discovering new technology opportunities: Keyword-based patent map approach. Technovation 29(6-7), 481–497 (2009)

[4] Campbell, R.S.: Patent trends as a technological forecasting tool. World Patent Information 5(3), 137–143 (1983)

[5] Lee, J.H., Kim, G.J., Park, S.S., Jang, D.S.: A study on the effect of firm's patent activity on business performance - Focus on time lag analysis of IT industry. Journal of the Korea Society of Digital Industry and Information Management 9(2), 121–137 (2013)

[6] Chen, Y.S., Chang, K.C.: The relationship between a firm's patent quality and its market value: The case of US pharmaceutical industry. Technological Forecasting and Social Change 77(1), 20–33 (2010)

[7] Lanjouw, J.O., Schankerman, M.: Patent quality and research productivity: Measuring innovation with multiple indicators. The Economic Journal 114(495), 441–465 (2004)

[8] Hair, J.F., Black, B., Babin, B., Anderson, R.E.: Multivariate Data Analysis (1992)

[9] Youk, Y.S., Kim, S.H., Joo, Y.H.: Intelligent data reduction algorithm for sensor network based fault diagnostic system. International Journal of Fuzzy Logic and Intelligent Systems 9(4), 301–308 (2009)

[10] Keum, J.S., Lee, H.S., Masafumi, H.: A novel speech/music discrimination using feature dimensionality reduction. International Journal of Fuzzy Logic and Intelligent Systems 10(1), 7–11 (2010)

[11] Jolliffe, I.T.: Principal Component Analysis (2002)

[12] Park, W.C.: Data mining concepts and techniques (2003)

[13] Uhm, D.H., Jun, S.H., Lee, S.J.: A classification method using data reduction. International Journal of Fuzzy Logic and Intelligent Systems 12(1), 1–5 (2012)

[14] Yi, J.H., Jung, W.K., Park, S.S., Jang, D.S.: The lag analysis on the impact of patents on profitability of firms in software industry at segment level. Journal of the Korea Society of Digital Industry and Information Management 8(2), 199–212 (2012)

[15] Organization for Economic Co-operation and Development, Economic Analysis and Statistics Division: OECD Science. Technology and Industry Scoreboard: Towards a Knowledge-based Economy (2001)

A Small World Network for Technological Relationship in Patent Analysis

Sunghae Jun* and Seung-Joo Lee

Department of Statistics, Cheongju University

Abstract. Small world network is a network consisted of nodes and their edges, and all nodes are connected by small steps of edge. Originally this network was to illustrate that the human relationship was connected within small paths. In addition, all people know and are related together within some connected edges. In this paper, we also think that the relationship between technologies is similar to the human connection based on small world network. So, we try to confirm the small world network in technologies of patent documents. A patent has a lot of information of developed technologies by inventors. In our experimental results, we show that small world network is existed in technological fields. This result will contribute to management of technology.

Keywords: Small World Network, Technological Relationship, Patent Analysis, Management of Technology.

1 Introduction

Small world network is a network consisted of nodes and their edges, and all nodes are connected by small steps of edges [1]. In general, a product is developed by combining detailed technologies, and diverse technologies have hierarchical structures. International patent classification (IPC) is a hierarchical system recognized officially over the world [2]. A patent document contains information about developed technologies [3], including the title, abstract, inventors, applied date, citation, and images [4]. The IPC code represents the hierarchical structure of technologies, which are classified to approximately 70,000 sub-technologies [2]. Many technologies are very different from each other; however, in this paper, we demonstrate that most technologies are closely connected either directly or indirectly, similar to the small-world network. We verify this similarity by simulation and experiment using real IPC codes and patent data. In our simulation, we will construct a social network analysis (SNA) graph [5, 6] and compute the connecting steps of the shortest path [7]. In the SNA graph, the node and the edge are the IPC code of the patent document and the direct connection or step between IPC codes, respectively. In the next section, we discuss the small-world network and six degrees of separation. Our proposed approach is in Section 3. The simulation and experimental results are shown in Section 4. In the last section, we conclude this research and present our plans for future work.

* Corresponding author.

K.M. Lee, S.-J. Park, J.-H. Lee (eds.), *Soft Computing in Big Data Processing*, Advances in Intelligent Systems and Computing 271,
DOI: 10.1007/978-3-319-05527-5_10,

2 Small World Network

The small-world network was originally researched in social psychology [8]. These studies found that most people (nodes) were connected directly or indirectly through their acquaintances by a limited number of steps (edges). This concept was extended in the "six degrees of separation" [7, 8] theory, which states that most people are connected by six or fewer steps [9]. In this study, we apply this theory to the network of technologies.

3 Technological Relationship by Small World Network

In this paper, we show that the network of technologies has the small-world network. This means that the small-world concept in social psychology can be applied to technology management, such as new product development and R&D planning. We performed a simulation and an experiment using real patent data. In the simulation, we generated a network of technologies randomly by selecting nodes and computing the shortest paths between any two nodes as follows:

Input1: Number of nodes
Input2: Probability of connections
Output1: SNA graph
Output2: Steps in shortest paths

Step1: Generating random data (node, path)
(1.1) determining number of nodes n
(1.2) selecting probability of connections α/n (α is constant)

Step2: Constructing SNA graph
(2.1) using generated random data
(2.2) creating a graph based on SNA

Step3: Computing shortest paths
(3.1) building distance matrix of all nodes
(3.2) calculating shortest distances between any two nodes

Step4: Combining results of Steps 2 and 3
(4.1) computing percentages of connecting steps
(4.2) constructing percentage table of shortest paths

In this simulation, we generated random data consisting of nodes and paths, where a node represents a specific technology and a path is the link between two technologies. Another input for simulation is the probability of the connection between the two technologies. Using these input values, we generated a complete SNA graph. Next, we computed the shortest paths between any two nodes, considering all possible paths, and we created a percentage table of the shortest paths. Even though the number of nodes is increased, the number of shortest steps does not increase in proportion to the number of nodes.

In next section, we discuss the simulation we used to verify the small-world network in the network of technologies. We also discuss our experiment where we analyzed real technology data using patent documents retrieved from patent databases throughout the world [10, 11]. Fig. 1 shows the experimental process of our case study.

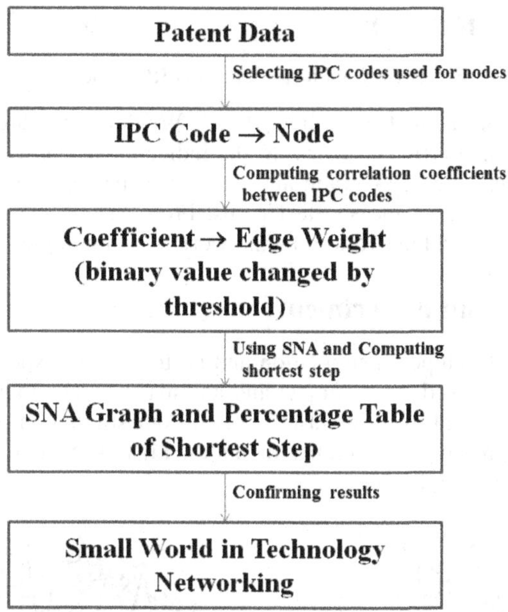

Fig. 1. Process for identifying connections in the network of technologies

First, we selected IPC codes from searched patent data. To construct the SNA graph, we computed the correlation coefficients between IPC codes. Fig. 2 shows the correlation coefficient matrix of IPC codes.

$$
\begin{array}{ccccc}
 & \mathbf{IPC_1} & \mathbf{IPC_2} & \cdots & \mathbf{IPC_n} \\
\mathbf{IPC_1} & \mathbf{1} & \mathbf{C_{1,2}} & \cdots & \mathbf{C_{1,n}} \\
\mathbf{IPC_2} & \mathbf{C_{2,1}} & \mathbf{1} & \cdots & \mathbf{C_{2,n}} \\
\vdots & \vdots & \vdots & \ddots & \vdots \\
\mathbf{IPC_n} & \mathbf{C_{n,1}} & \mathbf{C_{n,2}} & \cdots & \mathbf{1}
\end{array}
$$

Fig. 2. Correlation coefficient matrix of IPC codes

The correlation coefficient $C_{i,j}$ represents the strength of the relation between IPC_i and IPC_j [12]. The correlation coefficient of an IPC code with itself is 1. The value of a correlation coefficient is always between -1 and +1, inclusive. We transformed this value into binary data, which is the required format for input data in constructing the SNA graph. Fig. 3 shows the binary data structure, which becomes the "indicator matrix" in SNA [7].

$$\begin{array}{ccccc} & \mathbf{IPC_1} & \mathbf{IPC_2} & \cdots & \mathbf{IPC_n} \\ \mathbf{IPC_1} & \mathbf{0} & \mathbf{B_{1,2}} & \cdots & \mathbf{B_{1,n}} \\ \mathbf{IPC_2} & \mathbf{B_{2,1}} & \mathbf{0} & \cdots & \mathbf{B_{2,n}} \\ \vdots & \vdots & \vdots & \ddots & \vdots \\ \mathbf{IPC_n} & \mathbf{B_{n,1}} & \mathbf{B_{n,2}} & \cdots & \mathbf{0} \end{array}$$

Fig. 3. SNA indicator matrix of IPC codes

Each element $B_{i,j}$ of this matrix is either 0 or 1. We defined a threshold when converting the correlation coefficient matrix to the indicator matrix. Using the indicator matrix, we constructed the SNA graph and computed the shortest paths. Lastly, we built the percentage table of the shortest paths. From the result, we confirmed the existence of the small-world network in the network of technologies.

4 Simulation and Experiment

In this research, we developed a simulation and performed an experiment to confirm our theory. Further, we used the R-project and its "sna" package to perform computations in our simulation and experiment [13, 14]. In our simulation, we computed the shortest paths by increasing the number of nodes. Figs. 4, 5, and 6 show the SNA graphs by the number of nodes (n).

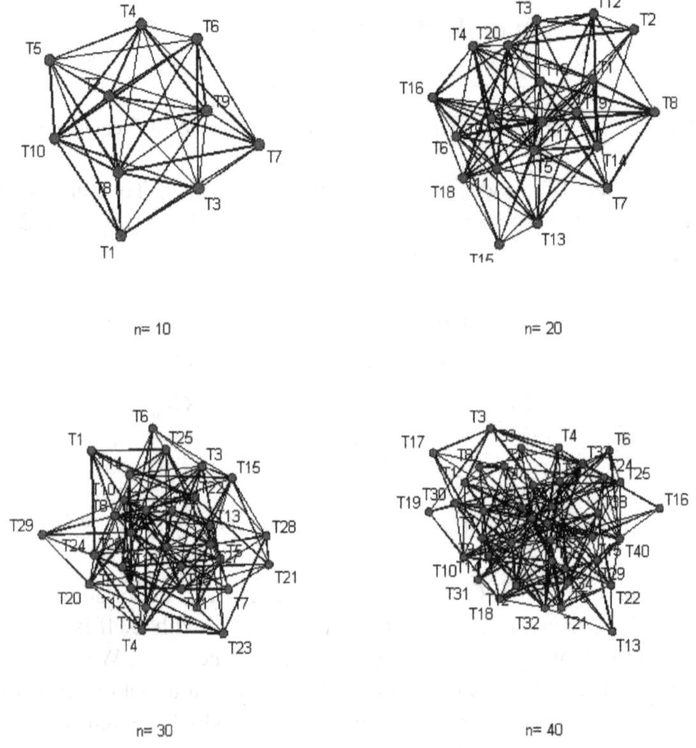

Fig. 4. Correlation graphs – simulation I: n=10, 20, 30, 40

Fig. 4 consists of the graphs for 10, 20, 30, and 40 nodes, where each node represents a specific technology. We determined that the nodes were connected to each other within five steps. The SNA graphs for 50, 60, 70, and 80 nodes are shown in Fig. 5.

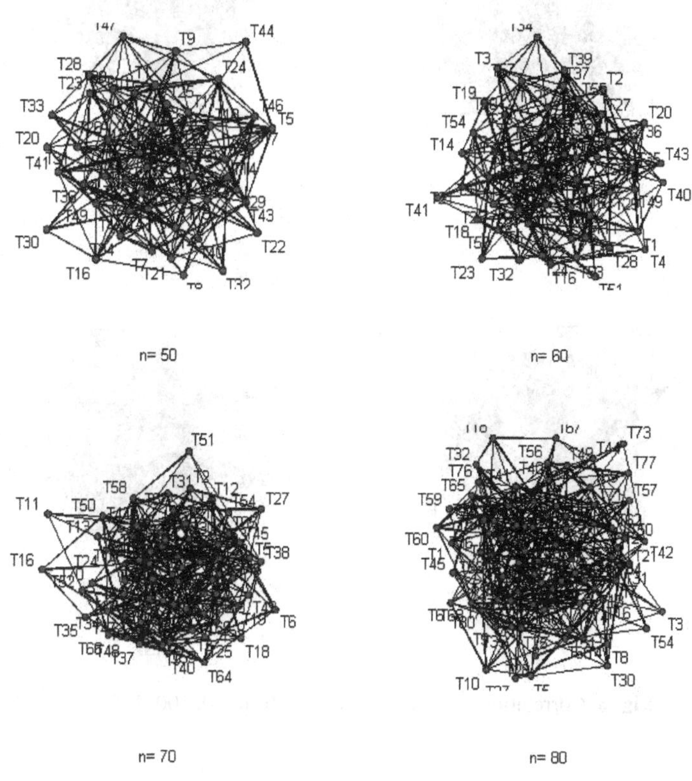

Fig. 5. Correlation graphs – simulation II: n=50, 60, 70, 80

Most steps are less than or equal to 5 in the simulations shown in Fig. 5. Fig. 6 represents the SNA graphs for 90, 100, 110, and 120 nodes.

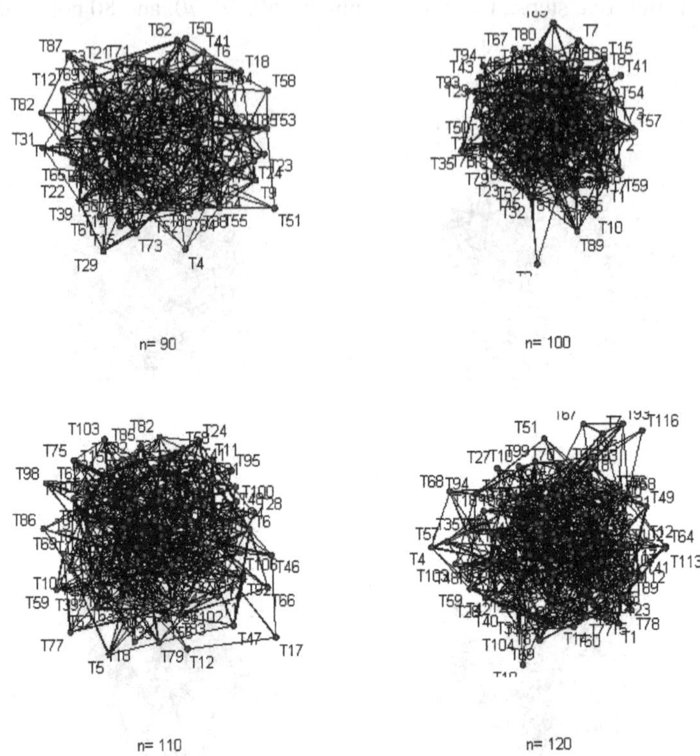

Fig. 6. Correlation graphs – simulation III: n=90, 100, 110, 120

In the simulation shown in Fig. 6, most connecting steps are less than or equal to six. More detailed results are shown in Table 1.

Table 1. Simulation results of shortest paths

Number of nodes	Shortest step (%)								
	1	2	3	4	5	6	7	8	Inf.
10	57.8	42.2	-	-	-	-	-	-	-
20	28.9	56.6	14.2	0.3	-	-	-	-	-
30	17.1	50.7	31.4	0.8	-	-	-	-	-
40	13.1	42.8	38.2	5.8	0.1	-	-	-	-
50	10.8	40.1	42.5	6.4	0.2	-	-	-	-
60	8.6	31.6	43.4	14.2	2.1	0.1	-	-	-
70	7.5	29.0	46.3	15.1	1.9	0.3	0.0	-	-
80	6.2	25.4	45.8	19.1	2.0	0.1	0.0	-	-
90	5.1	19.2	41.6	27.7	5.4	0.8	0.2	0.0	-
100	5.2	21.4	44.2	24.7	3.5	0.1			1.0
110	4.6	19.8	44.6	25.7	3.8	0.6	0.0	-	0.9
120	4.0	16.1	37.1	30.6	9.3	1.2	0.0	-	1.7

With 10 nodes, 46.4% of all paths between any two nodes are comprised of one step, and 48.2% are comprised of two steps. The remaining 5.5% are comprised of three steps. When the number of nodes is 70, 1.4% of all nodes are not connected, which is represented by infinite (Inf.); however, most paths are comprised of five or six steps. When the number of nodes is 120, most connecting steps are within five or six steps. Therefore, with this simulation, we confirmed that the small-world network exists in the network of technologies.

Using real patent data related to biotechnology, we performed an experiment to verify the small-world network in the network of technologies. The patent data consisted of 50,886 documents and 328 IPC codes. We used the IPC code as the node and randomly selected 10, 20, 30, and 40 codes from 328 IPC codes. Fig. 2 shows the SNA graphs for 10, 20, 30, and 40 nodes.

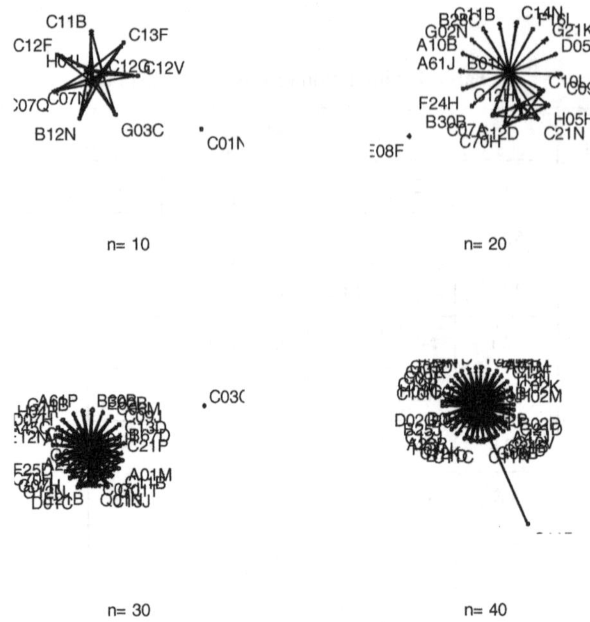

Fig. 7. Social graph – bio patent: n=10, 20, 30, 40

Each node represents the IPC code, and each path is indicated by a binary value in the indicator matrix. This value was determined from the correlation coefficient matrix based on a threshold. More detailed results are shown in Table 2.

Table 2. Experimental result of shortest path

Number of IPC code	Shortest step (%)			
	1	2	3	Inf.
10	61.8	38.2	-	-
20	23.8	76.2	-	-
30	36.3	63.7	-	-
40	36.2	63.8	-	-

When the number of IPC codes is 10, 61.8% of all paths are comprised of one step each, and the remaining 38.2% are comprised of two steps. With the increase of IPC codes, the percentage of two-step paths increased; however, the percentage of other connections between IPC codes did not change. Similar to the simulation result, this experiment shows that the increase of shortest path percentage is not proportional to the increase in number of IPC codes.

In this study, we confirmed the existence of the small-world network in the network of technologies by verifying that most connections between any two IPC codes were comprised of five or six steps, based on the results of our simulation and our experiment using real patent data. This information provides diverse strategies for technology management, such as R&D planning, new product development, and technological innovation.

5 Conclusions

In this research, we studied the existence of the small-world network in relationships between or the network of technologies. To confirm this network, we performed a simulation and an experiment using real patent data. We constructed SNA graphs and determined the shortest connecting paths between any two nodes. IPC codes were used as the nodes of the SNA graph, and a correlation coefficient matrix was used to determine the connections between nodes in SNA graph. From the results, we determined that an increase in the number of IPC codes did not affect the number of connecting steps between IPC codes. Most IPC codes were connected by five or six steps.

We conclude that, although most technologies may not seem related, they are, in fact, connected by technological association. We propose that this small-world network be used to manage technologies, such as new product development, R&D planning, and technological innovation. In our future work, we will verify the small-world network in diverse technology fields.

References

1. Watts, D.J., Strogatz, S.H.: Collective dynamics of 'small-world' networks. Nature 393, 440–442 (1998)
2. WIPO.: World Intellectual Property Organization (2013), `http://www.wipo.org`
3. Roper, A.T., Cunningham, S.W., Porter, A.L., Mason, T.W., Rossini, F.A., Banks, J.: Forecasting and Management of Technology. Wiley (2011)
4. Hunt, D., Nguyen, L., Rodgers, M.: Patent Searching Tools & Techniques. Wiley (2007)
5. Wasserman, S., Faust, K.: Social Network Analysis: Methods and Applications. Cambridge University Press (1994)
6. Pinheiro, C.A.R.: Social Network Analysis in Telecommunications. John Wiley & Sons (2011)
7. Huh, M.: Introduction to Social Network Analysis using R. Free Academy (2010)
8. Travers, J., Milgram, S.: An experimental Study of the small world problem. Sociometry 32(4), 425–443 (1969)
9. Han, J., Kamber, M.: Data Mining: Concepts and Techniques. Morgan Kaufmann (2005)
10. KIPRIS.: Korea Intellectual Property Rights Information Service (2013), `http://www.kipris.or.kr`
11. USPTO.: The United States Patent and Trademark Office (2013), `http://www.uspto.gov`
12. Ross, S.M.: Introduction to Probability and Statistics for Engineers and Scientists. Elsevier (2009)
13. Butts, C.T.: Tools for Social Network Analysis. Package 'sna', CRAN, R-Project (2013)
14. R Development Core Team.: R: A language and environment for statistical computing. R Foundation for Statistical Computing, Vienna, Austria (2013), `http://www.R-project.org` ISBN 3-900051-07-0

Key IPC Codes Extraction Using Classification and Regression Tree Structure

Seung-Joo Lee and Sunghae Jun*

Department of Statistics, Cheongju University

Abstract. Patent are results of researched and developed technologies. So we can get novel information for technology management by patent analysis. In the patent analysis, we use diverse items such as abstract and citation for discovering meaningful information. International patent classification (IPC) code is one of items for patent analysis. But, all IPC codes are needed to analyze patent data. In this paper, we propose a method for extracting key IPC codes representing entire patents. We use classification tree structure to construct our model. To verify the improved performance of proposed model, we illustrate a case study using retrieved patent data from patent databases in the world.

Keywords: IPC, Key IPC Code Extraction, Classification Tree Structure, Patent Analysis.

1 Introduction

Technology developments have been used to improve the quality of life in diverse fields. Science and technology are related to human society, so we need to understand and analyze technology. Patents are a good result of technology development [1], and patent documents include many descriptions of technology, including the patent title and abstract, inventors, technological drawing, claims, citation, applied and issued dates, international patent classification (IPC), and so forth [2]. To understand and analyze technology, patents are an important resource. Much research has focused on technology analysis in diverse domains [3-9]. In this paper, we analyze patent data for technology understanding and management. In particular, we use IPC codes extracted from patent data for patent analysis. We analyze IPC code data using a classification and regression tree (CRT), which is popular algorithm for data mining [10]. CRT induction consists of two approaches: classification and regression trees. Classification trees are used for binary or categorical data types of target variables, and regression trees are for continuous target variables. To use a CRT algorithm, we transform retrieved patent data into structured data for CRT input. Then, we preprocess retrieved patent documents to construct structured data. In this paper, we build IPC code data, where the columns and rows represent IPC codes and patents, respectively. Our patent analysis results show the relationships between technologies. Using these

* Corresponding author.

K.M. Lee, S.-J. Park, J.-H. Lee (eds.), *Soft Computing in Big Data Processing*, Advances in Intelligent Systems and Computing 271,
DOI: 10.1007/978-3-319-05527-5_11,

results, a company or a nation can create effective R&D policy and benefit from accurate technology management. To show our performance and application, we carry out an experiment using real patent data from international patent databases [11-13].

2 Research Background

A CRT is a tree structure based on nodes (leaves) and branches. Using a node and branch, a CRT describes the dependencies between variables of given data. A CRT used for data analysis involves two steps: tree construction and tree pruning. First, in tree construction, all data elements start in the root node. These are then divided into nodes by Chi square automatic interaction detection (CHAID), a Gini index, and Entropy [9]. Next, we remove the meaningless and more detailed branches to create a generalized CRT model. Fig. 1 shows a CRT model based on five input variables: A, B, C, D, and E.

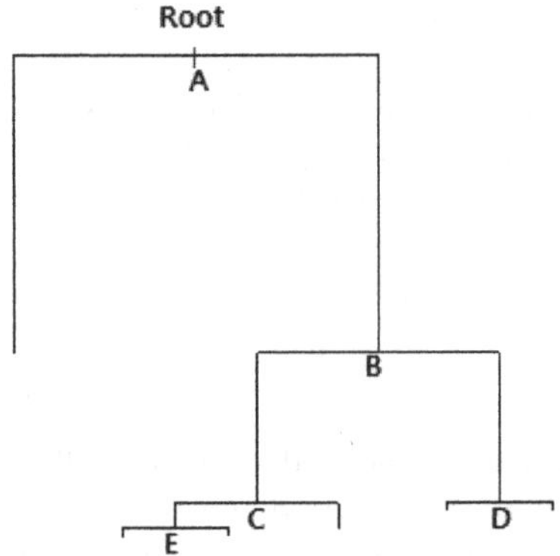

Fig. 1. Classification and regression tree structure

All the data elements are next divided into terminal nodes. A terminal node is an end node that cannot be further separated. The CRT structure in Fig. 1 has six terminal nodes. The elements of a terminal node are relatively homogeneous. In general, CRT output, the first used variable is more important to the target variable. For example, in Fig. 1, A is most important variable for the target variable. In other words, A has the biggest influence on the target variable. E has the smallest effect on the target variable.

3 A Classification and Regression Tree Model for IPC Data Analysis

We introduce a CRT model for IPC code data analysis. A patent has one or more IPC codes that describe technological characteristics. We can analyze technologies using IPC code data because an IPC code is a hierarchical system for defining technologies.

The top level of IPC codes consists of eight technological levels [14] (A, B, C, D, E, F, G, and H) that represent these technologies: "human necessities," "performing operations, transporting," "chemistry, metallurgy," "textiles, paper," "fixed constructions," "mechanical engineering, lighting, heating, weapons, blasting," "physics," and "electricity," respectively. All IPC code sublevels represent more detailed technologies. In this paper, we use IPC codes with high four digits, such as "F16H." Furthermore, we consider a CRT algorithm based on the Gini index for IPC code data analysis. When using the Gini index, we assume several candidate values for IPC code splitting. If an IPC code A has n elements, the Gini index for A is defined as follows:

$$Gini_A = 1 - \sum_{i=1}^{n} p_i^2$$

where p_i is the relative frequency of level i in A.

$$p_i = \frac{frequency\ of\ i\ levels\ in\ A}{n}$$

The IPC code with the smallest Gini index value is chosen to be split. We propose a full CRT model using all IPC codes and a reduced model that uses select IPC codes. For the reduced model, we compute the correlation coefficient as follows [15]:

$$\mathrm{Corr(A, B)} = \frac{Cov(A,B)}{\sqrt{Var(A)Var(B)}}$$

where Var(A) is the variance of IPC code A and Cov(A,B) is the covariance of IPC codes A and B. The correlation coefficient is a value between –1 and 1. To select the IPC codes for the reduced model, we use a p-value that is the smallest significance when the data rejects a null hypothesis [16]. The smaller the p-value, the more significant is the IPC code. For example, for a 95% confidence level, we decide an IPC code is significant when its p-value is less than 0.05. Therefore, we perform CRT analysis using full and reduced models for IPC code data analysis. The following process shows CRT generation for the full and reduced approaches.

(Step 1) Get patent documents for a target technology field.
(Step 2) Preprocess the patent data.
(Step 3) Extract the IPC codes.
(Step 4) Compute the correlation coefficients between the IPC codes and target technology. (This is an optional step for the reduced model.)
  (4.1) Select IPC codes to generate the CRT model by predetermined probability values, such as 0.05 or 0.001.
  (4.2) Use the selected IPC codes for the reduced model.
(Step 5) Generate the Classification and regression tree.
  (5.1) Assign all patents to a root node.
  (5.2) Divide each patent into the next node.
    (5.2.1) Use IPC code-selection measures.

(5.2.2) Increase the homogeneity of the node.
(5.2.3) Stop CRT generation when all nodes are terminals
(5.3) Remove some nodes.
(5.3.1) Use IPC code-pruning measures.
(5.3.2) Stop branch pruning when the CRT model is generalized.
(Step 6) Determine the important IPC codes for the target technology.

In this process, step 4 is optional when constructing a reduced model. This process shows our entire process from patent searching to determining important technologies for a target domain. The final result can be used for technology management such as R&D planning, new product development, and company or nation technological innovation. We carry out a case study using real patent documents in the next section.

4 Experimental Results

For our experiment, we searched patent data related to Hyundai from patent databases [11-13] to verify the performance of our model. We performed a case study to evaluate how our work is applied to real technology problem using an R-project and "tree" package [17, 18]. The retrieved patent data had 2,160 patent documents and 234 IPC codes. We used high 21 IPC codes with more than 100 frequencies, as shown in Table 1.

Table 1. IPC codes with more than 100 frequencies

IPC code	Frequency	IPC code	Frequency	IPC code	Frequency
F16H	1,897	B60W	250	B60J	153
B60R	542	F02B	245	B60N	130
B62D	518	F16F	200	F01N	128
B60G	479	F01L	194	G06F	118
F02D	464	H01M	187	C08L	112
B60K	422	B60T	164	F01P	106
F02M	343	F02F	158	F16D	104

F16H was the IPC code that most frequently appeared in the patent data, so we selected it as our target technology. The technology behind F16H is defined as "Gearing" by WIPO [6]. The rest of the IPC codes were used for exploratory variables in our model. Thus, our full model is shown as follows:

$$\text{F16H} = f_{DT}(B60R, B62D, \ldots, F01P, F16D)$$

f_{DT} represents a CRT model, and F16H is a response variable. The 20 variables of B60R, B62D, B60G, F02D, B60K, F02M, B60W, F02B, F16F, F01L, H01M, B60T, F02F, B60J, B60N, F01N, G06F, C08L, F01P, and F16D are exploratory variables. To construct the best model, we performed model selection from the full model to the reduced model. First, we used all the IPC codes to build our CRT model. Figure 2 shows a tree plot of the full model.

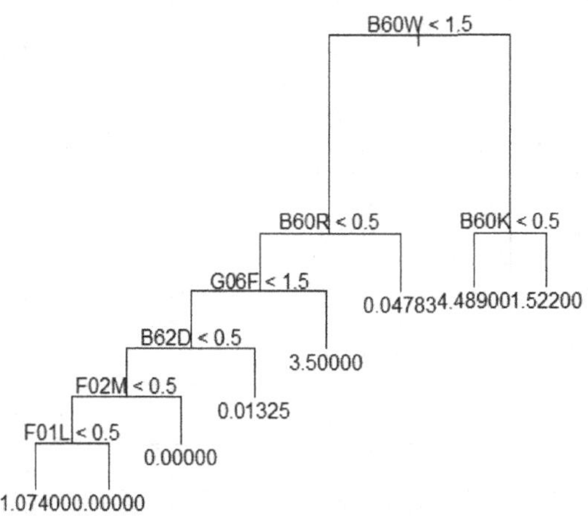

Fig. 2. Classification and regression tree diagram using full IPC codes

Seven IPC codes were selected for the constructed model: B60W, B60R, G06F, B62D, F02M, F01L, and B60K. Their defined technologies are "Conjoint control of vehicle sub-units of different type or different function; control systems specially adapted for hybrid vehicles; road vehicle drive control systems for purposes not related to the control of a particular sub-unit;" "Vehicles, vehicle fittings, or vehicle parts, not otherwise provided for;" "Electric digital data processing;" "Motor vehicles; trailers;" "Supplying combustion engines in general with combustible mixtures or constituents thereof;" "Cyclically operating values for machines or engines;" and "Arrangement or mounting of propulsion units or of transmissions in vehicles; arrangement or mounting of plural diverse prime-movers; auxiliary drives; instrumentation or dashboards for vehicles; arrangements in connection with cooling, air intake, gas exhaust, or fuel supply, of propulsion units, in vehicles," respectively. We knew that the technology of F16H was most affected by B60W technology because B60W was the first used for tree expanding. The technologies of B60R and B60K were the next most important to developing F16H technology. Next, the G06F, B62D, F02M, and F01L technologies affect F16H (in order of importance). The rest of the IPC codes did not affect technological development of F16H because they were not selected in the constructed tree model.

Next, we considered a reduced model without all the IPC codes. To select the IPC codes for use in our CRT model, we performed correlation analysis of F16H and all the input IPC codes. The correlation coefficient and p-value are shown in Table 2.

Table 2. Correlation analysis result of F16H and input IPC codes

IPC code	P-value	Correlation coefficient
B60R	0.0000	-1.1045
B62D	0.0000	-0.0900
B60G	0.0005	-0.0747
F02D	0.1359	-0.0321
B60K	0.0006	0.0742
F02M	0.0020	-0.0665
B60W	0.0000	0.2075
F02B	0.0247	-0.0483
F16F	0.0587	-0.0407
F01L	0.0013	-0.0693
H01M	0.0069	-0.0581
B60T	0.0526	-0.0417
F02F	0.0232	-0.0489
B60J	0.0023	-0.0656
B60N	0.0220	-0.0493
F01N	0.0468	-0.0428
G06F	0.0000	0.1055
C08L	0.1165	-0.0338
F01P	0.1740	-0.0293
F16D	0.2362	-0.0255

The B60R, B62D, B60W, and G06F IPC codes were strongly correlated with F16H because their *p*-values were small. Other IPC codes were selected by their *p*-values based on a predefined threshold. In this paper, we used two types of threshold values: 0.05 and 0.001. Thus, according to the *p*-values, we obtained different results for the CRT model. When the threshold for the *p*-value was 0.05, we selected the B60R, B62D, B60G, B60K, F02M, B60W, F02B, F01L, H01M, F02F, B60J, B60N, F01N, and G06F IPC codes to construct the tree model. The result of this reduced model was same as the full model. Next, we selected the B60R, B62D, B60G, B60K, B60W, and G06F IPC codes with *p*-values less than 0.001. The tree graph of this reduced model is shown in Figure 3.

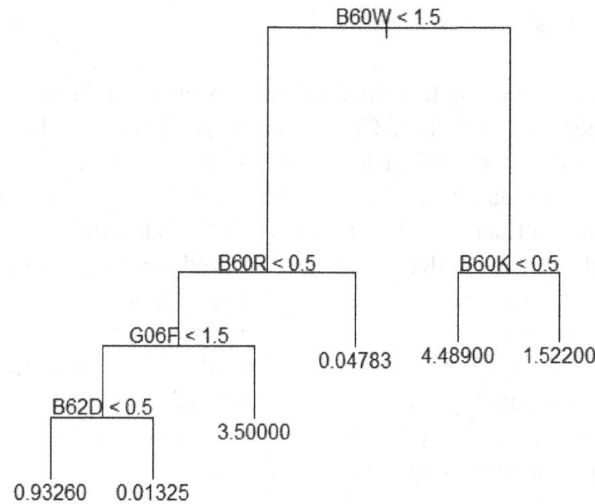

Fig. 3. Classification and regression tree diagram using reduced IPC codes

The reduced model in Figure 3 shows a part of tree graph for the full model. F02M and F01L were removed from the tree graph of full model. A more detailed comparison of the three models is shown in Table 3.

Table 3. Results of three Classification and regression tree models

Result	Full Model	Reduced Model (p-value=0.05)	Reduced Model (p-value=0.001)
Target code	F16H	F16H	F16H
Input codes	B60R, B62D, B60G, F02D, B60K, F02M, B60W, F02B, F16F, F01L, H01M, B60T, F02F, B60J, B60N, F01N, G06F, C08L, F01P, F16D	B60R, B62D, B60G, B60K, F02M, B60W, F02B, F01L, H01M, F02F, B60J, B60N, F01N, G06F	B60R, B62D, B60G, B60K, B60W, G06F
Actually used code	B60W, B60R, G06F, B62D, F02M, F01L, B60K	B60W, B60R, G06F, B62D, F02M, F01L, B60K	B60W, B60R, G06F, B62D, B60K
Number of terminal node	8	8	6
Residual mean deviance	4.352	4.352	4.451

The target IPC code for all three models was F16H, but the input IPC codes were determined by the various models. We compared the residual mean deviance (RMD) for the three models. The smaller the RMD, the better constructed the model [18]. We found that the full model and reduced model with a *p*-value of 0.05 were better than the reduced model with a *p*-value of 0.001.

5 Conclusions

Classification and regression tree models have been used in diverse data mining works. In this paper, we applied the CRT model to patent data analysis for technology forecasting. Using IPC codes extracted from patent documents, we performed CRT modeling for IPC code data analysis. We selected target and exploratory IPC codes and determined the influences of exploratory IPC codes on the target IPC code. We constructed three tree models using full and reduced approaches. Based on the results of the tree models, we identified important technologies affecting the target technology as well as technological relationships between the target and exploratory technologies. Our research contributes meaningful results to company and nation R&D planning and technology management. In future work, we will study a more advanced tree model, including classification and regression trees, and develop a new product development model using the tree structure.

References

1. Roper, A.T., Cunningham, S.W., Porter, A.L., Mason, T.W., Rossini, F.A., Banks, J.: Forecasting and Management of Technology. Wiley (2011)
2. Hunt, D., Nguyen, L., Rodgers, M.: Patent Searching Tools & Techniques. Wiley (2007)
3. Jun, S.: Central technology forecasting using social network analysis. In: Kim, T.-H., Ramos, C., Kim, H.-K., Kiumi, A., Mohammed, S., Ślęzak, D. (eds.) ASEA/DRBC 2012. CCIS, vol. 340, pp. 1–8. Springer, Heidelberg (2012)
4. Jun, S.: Technology network model using bipartite social network analysis. In: Kim, T.-H., Ramos, C., Kim, H.-K., Kiumi, A., Mohammed, S., Ślęzak, D. (eds.) ASEA/DRBC 2012. CCIS, vol. 340, pp. 28–35. Springer, Heidelberg (2012)
5. Jun, S.: IPC Code Analysis of Patent Documents Using Association Rules and Maps-Patent Analysis of Database Technology. In: Kim, T.-H., et al. (eds.) DTA / BSBT 2011. CCIS, vol. 258, pp. 21–30. Springer, Heidelberg (2011)
6. Jun, S.: A Forecasting Model for Technological Trend Using Unsupervised Learning. In: Kim, T.-H., et al. (eds.) DTA / BSBT 2011. CCIS, vol. 258, pp. 51–60. Springer, Heidelberg (2011)
7. Fattori, M., Pedrazzi, G., Turra, R.: Text mining applied to patent mapping: a practical business case. World Patent Information 25, 335–342 (2003)
8. Indukuri, K.V., Mirajkar, P., Sureka, A.: An Algorithm for Classifying Articles and Patent Documents Using Link Structure. In: Proceedings of International Conference on Web-Age Information Management, pp. 203–210 (2008)
9. Kasravi, K., Risov, M.: Patent Mining - Discovery of Business Value from Patent Repositories. In: Proceedings of 40th Annual Hawaii International Conference on System Sciences, pp. 54–54 (2007)
10. Han, J., Kamber, M.: Data Mining: Concepts and Techniques. Morgan Kaufmann (2005)
11. KIPRIS.: Korea Intellectual Property Rights Information Service (2013), http://www.kipris.or.kr
12. USPTO.: The United States Patent and Trademark Office (2013), http://www.uspto.gov
13. WIPSON.: WIPS Co., Ltd. (2013), http://www.vipson.com
14. WIPO.: World Intellectual Property Organization (2013), http://www.wipo.org

15. Ross, S.M.: Introduction to Probability and Statistics for Engineers and Scientists. Elsevier (2009)
16. Ross, S.M.: Introductory Statistics. McGraw-Hill (1996)
17. R Development Core Team.: R: A language and environment for statistical computing. R Foundation for Statistical Computing, Vienna, Austria (2013), `http://www.R-project.org` ISBN 3-900051-07-0
18. Ripley, B.: Classification and regression trees, Package 'tree'. R Project – CRAN (2013)

Author Index